TABLE O

CHAPTER ONE
    Introduction 1

CHAPTER TWO
    Location 5

CHAPTER THREE
    Wrangellia 9

CHAPTER FOUR
    Geology and Gold Mineralization 13

CHAPTER FIVE
    The Last Ice Age 19

CHAPTER SIX
    First Europeans Enter the Nabesna Area 23

CHAPTER SEVEN
    Gold Rush Era Early Prospectors Enter the Nabesna Area 29

CHAPTER EIGHT
    The Shushana Gold Rush: A Prelude to the Nabesna Gold Mine . .35

CHAPTER NINE
    Roads, Schools and People 43

CHAPTER TEN
    The Bear Vein: The discovery that made Nabesna Gold Mine 51

CHAPTER ELEVEN
    The beginning of the Nabesna Gold Mine 61

CHAPTER TWELVE
    Litigation 65

TABLE OF CONTENTS (CONT.)

CHAPTER THIRTEEN
    Building the Nabesna Gold Mine 71

CHAPTER FOURTEEN
    Milling Ore and the Recovery of Gold 91

CHAPTER FIFTEEN
    Nabesna Gold Mine Town 101

CHAPTER SIXTEEN
    World War II and the Final Days of the Nabesna Gold Mine 119

CHAPTER SEVENTEEN
    The Early Cold War Period 129

CHAPTER EIGHTEEN
    Conclusion 135

To John & EA Patterson

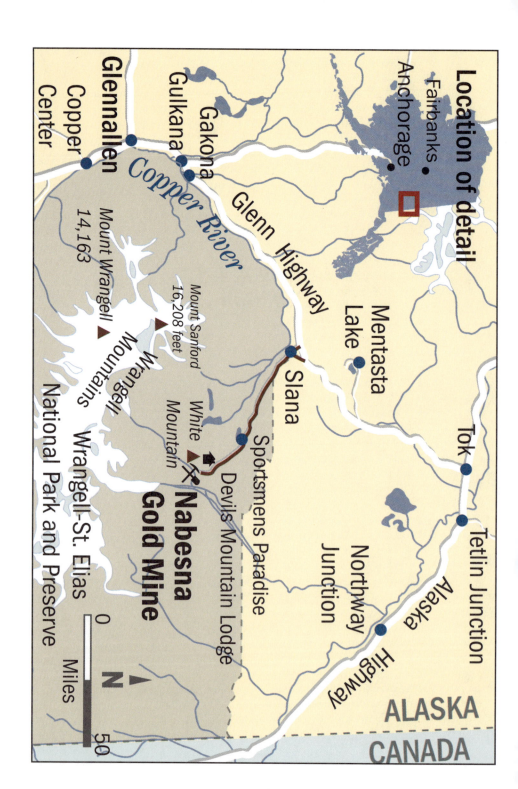

CHAPTER 1

Introduction

Long ago when Alaska was a Territory, the great eastern banks financed, and large mining concerns ran, the well-known gold and copper mines in Alaska. Nabesna Gold Mine was different. Carl F. Whitham, almost single-handedly, built Nabesna Mine. Whitham was 18 years of age in 1900 when he arrived in Alaska. He had no formal education. Whitham nevertheless built Nabesna into the largest hard rock gold mine in eastern interior Alaska, and he did it beyond the frontier of the Territory of Alaska without the aid and control of any large financial institution.

Nabesna Gold Mine, closed now for more than a half-century as a producing gold mine was the fourth largest gold producer during the Territorial days of Alaska. That production record stood all the way to 1983. But that fact standing alone reveals little of the worth and significance of this mine to eastern Alaska. No doubt much of the value of the mine is indiscernible today, but long ago it opened up that part of Alaska and during the Great Depression provided desperate men jobs where none existed before. But the benefits were more than that.

Without Nabesna Mine, there would be no Nabesna Road, a road built by Carl Whitham to reach and supply the mine. Without that road there would be no Reeve Field. That field, originally called Nabesna Landing Field, and built by the mine to fly gold out, made possible the Civil Aeronautics Authority's facilitating the world's largest airlift to 1941. That airlift allowed building of Northway airfield in 1941. Northway airfield was critical to ferrying lend lease warplanes to Russia in its desperate fight against the Nazis in World War II.

Without Nabesna Road and the airlift it made possible, Northway Airfield and Northway, a place often in winter the coldest in Alaska and the nation, might not exist. Without the mine and Nabesna Road, the Slana-Tok Cutoff, a major Alaska highway, would, at least as likely as it was built, not have been built.[1] Without Nabesna Road and the "Cutoff", the town of Tok might not be where it is to greet and serve persons traveling the Alaska Highway. Without Nabesna Road there would be no "northern gateway" to the Wrangell-St. Elias National Park and Preserve. There would necessarily be none of the settlement along that road. In fact, without the road, the park and preserve would undoubtedly not be of its present dimensions. Other effects of this mine on eastern Alaska are less clear; however, without that mine, nothing other than the remnants of the ancient Indian settlements of that area would be today's Nabesna.

Questions linger on at that old ghost town and mine. Long ago did the nearby war-like Batzulneta Indians, while hunting those cliffs on White Mountain, first discover one of the richest gold veins reported in Alaska? Did that gold-rich vein, called the "Bear Vein," the discovery of which allowed Carl Whitham to open his mine, end abruptly, or did a survey mistake deep underground in White Mountain lead Whitham's miners in the wrong direction and leave this rich vein there unworked and intriguing? Did the U.S. Military cache arms at Nabesna during the Cold War as several newspapers reported? Are some still there? While questions will endure, that old town and mine will remain a remembrance of a romantic, by-gone, era.

---

[1] The U.S. Army during World War II in 1942 started at Nabesna Road where it ran past the "Slana Roadhouse" and pushed a cut-off through the mountains to join with the newly built Alaska Highway. The town of Tok grew up at that junction. Thus explains the name "Slana-Tok Cutoff." After Statehood the "Cutoff" officially became part of the Glenn Highway. The Glenn Highway was named for Capt E.F. Glenn, U.S. Army, in honor of his early explorations in Alaska.

Carl F. Whitham came to Alaska in 1900 at age 18. He worked in various mines including his own gold placer mine at Chisana in the eastern Alaska Range. In 1926 he discovered one of the richest gold bearing veins reported in Alaska and founded Nabesna Mining Corporation. It became the largest hardrock gold mine in eastern Alaska and the 4th largest hardrock gold producer in Alaska.

White Mountain cliffs are said to be called "El Se Ba" by the old Batzulneta Indians. Nabesna Gold Mine town and mill at left below cliffs. Aerial tramways lowered gold ore from mine workings in the cliffs to the mill. Nabesna Road winds along the base to the old town. Lt. Allen in 1885 was the first European to report the Batzulnetas, who had guided him to the cliffs.

CHAPTER 2

Location

Nabesna Road heads off into the wilds in an easterly direction from Slana at Mile Post 60 of the Glenn Highway, locally called the "Slana-Tok Cutoff." Nabesna Road ends at Nabesna Gold Mine and town, 46 miles after it starts. The mine and its old town are as far as a person can drive on that road and about as far as anyone can drive an ordinary road vehicle into the Alaska wilderness anywhere. A lady journalist once described Nabesna Road as a "precarious trail to nowhere."[2] Some who have experienced that road would consider the journalist's assessment unintended praise.

The road ends at the old town and mine. There limestone cliffs of White Mountain tower above the town and have for hundreds, if not thousands, of years been a landmark for people throughout Nabesna Valley. From when the old Batzulnetas Indians were the undisputed sovereigns of the Nabesna, they called those cliffs "El Se Ba"[3,4]

Atop those cliffs a person looks down on the weathered buildings of the old mining town. A 2,000-foot long aerial tram from the mill stretches directly up at him or her. Outward from the cliffs stretches Nabesna Valley. In the foreground are Raven Hills and the gorge of Jacksina Creek, a "creek" whose meandering braids spread nearly a mile wide before it empties into the Nabesna River. Beyond that junction rises the Eastern Alaska Range and snow-capped Keenan Peak.[5] Farther up the valley, one sees the headwaters of Nabesna River and where it roars out from Nabesna Glacier to carry its load of glacial silt to the great rivers of central Alaska. Some say that two-mile wide 80-mile long glacier is the longest alpine glacier in the world. From those same cliffs one can see Orange Hill, where the glacier ends and the river begins. The name Orange Hill comes about by the tens of square miles of orange-yellow pyrite stained mountain slopes alongside that great

NABESNA GOLD

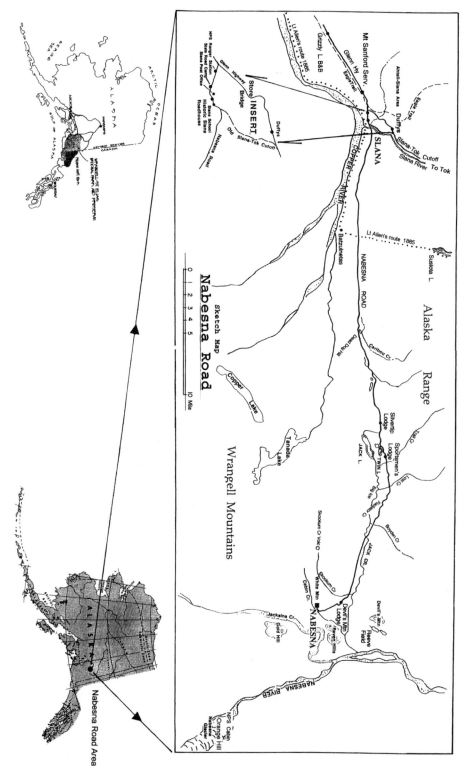

river of ice. The U.S. Geological Survey once called Orange Hill one of the largest undeveloped copper deposits in North America. In that huge iron stained and mineralized area the National Park Service in 2000 built a cabin for park visitors reached only on foot or by airplane. Those cliffs above the old mine town provide a panorama of the snow capped Eastern Alaska Range, the volcanic Wrangell Mountains and the eastern and southern part of the Wrangell-St. Elias National Park and Preserve.

The mine, with its portals in those cliffs on White Mountain, is on the National Register of Historic Places and has been so for a third of a century. There's no bronze plaque saying so. This designation is superfluous. By stepping into the old place a person knows he has stepped back into Alaska's history

The mine according to the best information goes back to 1898, but some of those workings look even older than that. In the years that have passed since the mine's beginnings, storm, fire, some vandalism and just plain old age have taken a toll on the buildings.

The mine is surrounded by the "Preserve" part of the Wrangell-St. Elias National Park and Preserve created in 1980. That's neither good nor bad. It was there long before the park and even before the area was mapped.

---

[2] Helen Gillette, Staff Writer, *Anchorage Times*, July 8, 1979.
[3] Name is stated in 1938 Quarterly Report, Nabesna Mining Corp. In Apache the word is said to mean something high and white. Lt. Allen in 1885 wrote many Batzulnetas Indian words were similar to Apache.
[4] Meyer, M.P., 2000, *Mineral assessment of Ahtna, Inc. selections in Wrangell-St. Elias National Park and Preserve, Alaska*, Final Report No. 34. Dept. of the Interior.
[5] F.J. "Joe" Keenan a lawyer by profession served as Director of Alaska Division of Lands after Alaska statehood in the sixties and seventies. Like his friend, Bob Atwood, publisher of the *Anchorage Times*, Joe Keenan was a staunch supporter of statehood and development of Alaska resources. As director, Joe Keenan conducted the first lease sale that became the great Prudhoe Bay oil field.

Looking south from Nabesna Gold Mine across Jacksina Creek and Nabesna River to the snow capped Eastern Alaska Range. Nabesna Glacier, the world's longest alpine glacier is to the extreme right and hidden by the hills. National Park Service airlifted a cabin to a site called Orange Hill for visitor use where the glacier ends and Nabesna River begins.

CHAPTER 3

Wrangellia

This land, this Nabesna, and in fact all the Wrangell Mountains of which it is a part, sits atop a drifting section of the earth's crust called a tectonic plate. Geologists call the plate "Wrangellia." Few people looking at today's Wrangellia with those towering snow-covered mountains would guess the circumstances of its modest dawning.

This dawning of Wrangellia was about 250 million years ago in a tropical sea thousands of miles from what we now call Alaska. Much of the material that makes up today's White Mountain came from that long ago, far away, place; still, there is a disconcerting mix of continuity and discontinuity between now and then in Wrangellia. Shark species, or their close cousins that survive today, already inhabited the sea of its beginning. Much of our genetic makeup was already to be found where and when Wrangellia formed. However, nothing about the land or where it was at its beginning would be recognizable today.

Hundreds of millions of years ago on the third planet from an inconsequential star, great lava flows spread across an unknown land. The land formed of those lava flows, flows that are now called greenstone at Nabesna, eventually sank beneath an ancient Triassic sea. During the following millions of years, limestone accumulated atop the greenstone. Then that ancient sea floor slowly rose so that a shallow gulf washed a Florida-sized landmass of limestone that stretched beyond the horizon.

Along that marshy shore animal life flourished and died for countless generations leaving a brown, limey mud. This limey mud thickened to a hundred feet or more and formed an algal mat. "Sabkha" is its more exotic name.[6]

This formation with its brown algal mat atop it again slowly sank below the sea, and, as tens of millions of years passed, more thousands of feet of limestone buried the algal mat. Later, during the Cretaceous period, thousands of feet of sea mud in turn covered the limestone. Gradually this small part of the earth's crust drifted from its beginning in the mysterious way of tectonic plates.

Wrangellia is one of the smaller tectonic plates. That means simply it is not one of the nine to twelve, depending on which expert is talking, large plates. For whatever reason, the Wrangellia plate drifted northward past what is now the West Coast of the United States and British Columbia. Inexorably it moved along and in doing so grounded here and there along that coast losing parts of itself wherever it made contact with the larger plate. Then, a relatively short time ago in geologic terms, around fifty million years, after the extinction so some say, of the last great dinosaurs, Wrangellia docked against what is now Alaska.

Back then, what we call Alaska today did not have its present-day outlines. While no one can say with certainty what this prehistoric land's contours were, geologists can determine where those large pieces, plates, of the earth's crust that formed present-day Alaska came together. The northeastern edge of Wrangellia is the Eastern Alaska Range. There the Denali Fault, only a few miles east of Nabesna Mine, separates Wrangellia from plates to the north. The traveler along the Slana-Tok Cutoff crosses the Denali Fault in Mentasta Pass in the Eastern Alaska Range.

After Wrangellia docked, heavier rocks of another plate then located in what became the Gulf of Alaska pushed to the northeast and under the lighter rocks of Wrangellia. Geologists call this process subduction.

As the subducted plate sank deeper through the crust of the earth into the mantle, it began to melt. As the melting continued, a catastrophic event deep below Nabesna was in the making. Great masses of molten granitic rock rose and invaded the limestone, and then millions of years later lava erupted and formed the great volcanoes. All the while the subducted plate acted like a wedge under Wrangellia and slowly uplifted the land. In time the uplift accelerated erosion, and White Mountain began to take shape. The toe of the limestone cliffs eroded, and the algal mat came into the light. For the first time in nearly 250 million years the algal mat saw the sun, the same sun that was essential to the life that formed it.

The algal mat's accessibility right at the base of the mountain and at one of the mine's portals provides an unusual opportunity. It allows a person to place a hand on a spot of rock and know it formed from life, much of which would be unrecognizable today, some 250 million years ago, at some far-off tropical place thousands of miles from where he or she is standing.

Nabesna Mine and the Wrangell Mountains sit atop a tectonic plate called Wrangellia. Some believe it drifted in from a distant South Pacific place and docked with what is now Alaska about 50 million years ago.

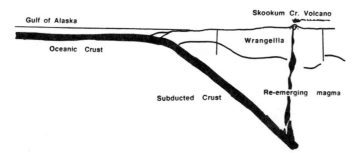

Rocks of what is now the Gulf of Alaska were pushed, subducted, under Wrangellia and pushed deeper into the mantle where the rocks melted. The melted rock, magma, rose through faults and two to three million years ago it erupted as lava and built Skookum Creek Volcano a few miles from what is now Nabesna Mine. Hundreds of feet of lava covered the area. In the millions of years since, erosion has stripped much of the lava away.

The dark bed is an algal mat at the toe of limestone cliffs near the 650 mine portal. The algal mat formed nearly 250 million years ago in a shallow sea somewhere in the South Pacific at the beginning of Wrangellia. Persons collecting raw gold eroded from the Bear Vein higher up on the cliffs built the ladder years ago.

---

[6] "Subkha" is said to be a term borrowed from the great expanse of algal mat found in the Persian Gulf.

CHAPTER 4

Geology and Gold Mineralization

The story of the Nabesna Mine is driven by gold. Better-accepted science holds that the algal mat has nothing to do with gold at Nabesna. However, other rocks visible on White Mountain and virtually next to the algal mat played essential roles in the gold mineralization.

Millions of years ago near the end of the Cretaceous or the early Tertiary period a large mass of molten granitic rock called Nabesna stock, as much as a mile or so wide and several miles long, intruded the limestone laid down in that faraway sea. At the onset of this intrusion, many of the rocks we see today as White Mountain remained buried thousands of feet below what was then the surface. The intrusion and later events changed all that.

The intrusion of the molten "Nabesna stock" recrystallized and hardened much of the limestone now seen on White Mountain and some was converted to a low-grade marble. The intrusion had some effect, perhaps a major effect, on altering the sea mud that long ago deposited above the limestone to a hard, fine grain, rusty colored rock called argillite.

The Nabesna stock is considered the major source of the valuable minerals at Nabesna. The earliest and hottest period of mineralization at Nabesna is called the "skarn" stage. Hot solutions containing exotic minerals escaped from the molten Nabesna stock. Those solutions reacted according to the laws of physics and chemistry, and the first solutions escaping from the stock were rich in iron. They migrated into the cooler limestone and/or low-grade marble, forming large deposits of coal black, high quality iron ore called magnetite, once called lodestone. Silica-rich solutions from the stock followed and altered the limestone in places to a sparkling white wollastonite.[7] Colorful garnets[8] and other semi-precious minerals formed later. Garnet of various colors makes up much of the rock seen above the old

White Mountain seen from Gold Hill. "QT" Lava covered top of mountain when Skookum Creek Volcano erupted 2 to 3 million years ago. "K" Cretaceous argillite. "Tr" Cliff forming Triassic limestone. "di" Diorite stock called Nabesna stock and one likely source of gold. "G" algal mate. "Trd" Dolomite. Nabesna Mill is the white dot at the base of the cliffs. Lines show aerial tram cables up cliffs. Nabesna Road leads around the mountain from the right to the old town. In far background is the Eastern Alaska Range. In foreground is Jacksina Creek. Cabin Creek runs from left to right past the town.

town. The last solutions expelled from the cooling Nabesna stock, maybe half a million or more years after the stock intruded the limestone, were rich in heavy metals including gold, silver and copper. This marked the end of the skarn stage. However, tens of millions of years later another mineralization event would occur unrelated to Nabesna stock.

Cooling during the high temperature skarn stage occurred slowly, maybe lasting thousands of years. Crystals had time to grow large. Gold and silver in the superheated solutions of the skarn stage combined with pyrite to form auriferous pyrite. Auriferous pyrite is a coarse crystalline mineral. In some areas of the mountain it replaced large areas of limestone and elsewhere it solidified as veins of rich ore. Auriferous pyrite is often intergrown with large translucent calcite crystals.

Over eons the land gradually uplifted. By 3 million years ago normal forces of erosion had probably shaped the land to look much as it does today–except for the lava flows. Those flows came about when Skookum

Looking west across Raven Hills and Jacksina Creek to rim of Skookum Creek caldera shown as dash line. "X" is Skookum Creek Volcano that erupted 2 to 3 million years ago. 1. Nabesna town and mill with underground mine workings above on White Mountain. 2. Cabin Creek. 3. White Mountain. 4. Nabesna Road. 5. Devils Mountain Lodge. 6. Jack Creek. 7. Skookum Creek. 8. Raven Hills. 9. Jacksina Creek. 10. Mt. Sanford. 11. Nabesna Road leads 46 miles to Slana and Glenn Highway. (Photo courtesy of USGS, 1940.)

Creek volcano erupted two to three million years ago.[9]

That Volcano is at the area of the head of Cabin and Skookum Creeks. Cabin Creek runs underground right along side and south of the old town. It was the source of water when the mine was operational and is the source of water at the old town today. Skookum Creek runs down the back side of the mountain and crosses Nabesna Road, sometimes in a boulder rolling rush, about five miles from the old town.

The main vent of the volcano is about 5 miles from and generally west of Nabesna Mine and the old town. The mass of White Mountain is thus between the old town and the now dormant Skookum Creek Volcano. However, the main vent is not the only vent. In fact, one secondary vent or cinder cone is atop White Mountain less than a mile from the old town. A hiker can see rusty pipes of others on the walls of Cabin Creek, just south and west of town.

Some believe the great Wrangell Lava Field that now stretches across thousands of square miles of eastern interior Alaska first started with the eruption of Skookum Creek Volcano. Then after changing the face of Nabesna, the volcanism with its lava flows migrated to the west, forming the starkly volcanic mountains like Sanford and Drum and others so visible from the Slana-Tok Cutoff.

In the next half million years or so, after the volcano's first eruption, billions of tons of lava poured out of Skookum Creek Volcano and other vents. At least 600 feet of lava covered White Mountain. Most of it is still up there. Elsewhere in the lower areas it piled up thousands of feet thick. The air must have reeked with choking fumes. Acid waters of every description circulated through and altered those steaming lava beds and the rocks below them. Nabesna was a nasty place back then.

In time the huge magma chamber, the source of the lava, deep below Skookum Creek Volcano emptied and a great volume of lava that erupted from the chamber piled-up on top of its roof. When the roof, and it was tens of square mile in area, could no longer support that great weight it collapsed. The fractured and broken rock slid down the sides as it might a huge bowl to fill the chamber. Skookum Creek caldera formed. The caldera, according to the USGS (1995), is at least 10 miles long by 4 miles wide.[10] Crater Lake in Oregon is also a caldera.

Nabesna Mine is on the eastern rim of the caldera, and rocks there are faulted, fractured and crushed. After the caldera collapsed, surface waters mingled with hot solutions of volcanic origin and became acid. As those chemically charged corrosive solutions circulated, gold, silver and other minerals that deposited during the early skarn stage leached from the broken and crushed rocks. The leaching process took place relatively near the surface, and the circulating solutions might have been a little hotter than steamy dishwater. Crystals of metals and minerals leached from the broken rocks and carried in solutions within this cooler environment never had the chance to grow large. Therefore, the metals and minerals precipitated as fine grain crystals. Gold deposited from these cooler late solutions is found with fine grain, vuggy and sometimes ribbon quartz. Ribbon quartz demonstrates repeated deposition along a vein or lode. Calcite, so common with the much older skarn ore, is rare in the late-formed mineralization. The two types of mineralization, although deposited millions of years apart, are sometimes found side by side in the mine.

The largest tonnage of gold and silver ore mined at Nabesna was from the skarn deposits. This ore graded about one ounce gold per ton of ore and one to two ounces silver per ton. The late-formed gold bearing quartz ore was much higher grade. The highest grades of ore deposited during that late stage, reported to date, exceeded several hundred ounces of gold per ton and more in silver.

When the two types of gold ore are held together, the differences are apparent. The early deposited skarn ore, the auriferous or gold-pyrite ore, is brassy yellow and coarsely crystalline; whereas the late deposited fine-grained gold-quartz ore is glassy and translucent, with a milky or bluish tint. The quartz fractures with razor sharp edges. This was a feature the resourceful Batzulnetas may have found very useful.

---

[7] Wollastonite is a calcium silicate formed under high temperature. Since 1990 its value has much increased for use in plastics and as a replacement for asbestos.

[8] Garnet is not one mineral but is the name of a group of 14 related silicate minerals and all are semi-precious gems. It is the difference in metal elements that give each individual garnet its different color. At Nabesna the dark red garnets are the most common. The word "garnet" alludes to the deep red color and comes from the Latin "granatum," meaning pomegranate, a fruit characterized by its red seeds. Garnets are gemstones and the most rare and most expensive is the green garnet, "demanitoid." The Egyptians fashioned garnets into beads before 3100 B.C. Some believed garnets possessed spiritual power and a fire glowed within them. Believing that spiritual power, the Hunzukut are said to have used garnet bullets to fight the British in the Kashmir. The Talmud told that Noah used a single garnet to light his entire Ark. There is a Sherlock Holmes mystery titled, "The Adventure of the Blue Carbuncle." Carbuncle in his time also referred to garnet. The keen Holmes makes his only mistake... garnets come in many colors but never blue.

[9] Richter, Smith, Lanphere, Dalrymple & Shew (199), "Age and Progression of volcanism, Wrangell volcanic field, Alaska." Bulletin of Volcanology. Spring 1990. Also see Richter (1995) "Guide to the Volcanoes of the Western Wrangell Mountains, Alaska-Wrangell-St. Elias National Park and Preserve."

[10] Because of its unique geology, several universities have used Nabesna (since World War II) as a geology field camp. Not all of those researches agree that a caldera formed.

CHAPTER 5

The Last Ice Age

The snow and ice fields of the Wrangell Mountains now spread like a sparkling blanket of white to the far horizon. Tranquility is the rule. They are interspersed with forests and punctuated with rivers and lakes. It was very different several million years ago. The Wrangells must have been a hellish place when those great volcanic mountains were building and billions of tons of lava poured out and changed the landscape forever. Great clouds of volcanic ash from these interior volcanoes darkened the skies and more was added to the northern skies by volcanoes erupting along the Alaska Peninsula, the Aleutian Chain and the Kamchatka Peninsula in what is now Siberia. Winds blowing out of Siberia pushed those ash clouds across Alaska, Canada and into northern Europe.[11]

Very late in this time of geologic turmoil the hellishness became one of cold, as opposed to the heat and volatility of the earlier times. The Arctic Ocean froze and the polar ice cap formed. These events coincided with a time when the earth reached its most outward orbit from the sun, supposedly every 70,000 to 100,000 years or so. The air grew even colder and the last great ice age began.[12]

Where travelers today on Nabesna Road see splendid natural diversity they would, if they could warp back 15,000 years, see a great wall of ice in Nabesna Valley. They would be experiencing the ubiquity and vastness of the ice of the last great ice age. They would see in the valley from their temporal vantage point 15,000 years ago little else besides the ancestral Nabesna-Jacksina Glacier. That river of ice, flowing out of the Wrangell Mountains, spread across Nabesna Valley, sculpted the lower slopes of the limestone cliffs of White Mountain, gouged out chunks of the algal mat, and continued eastward through the gap in the Alaska Range. From there this tremendous river of ice finally spilled into the Tanana River valley.

During that last great ice age fresh water became part of the great continental ice sheets. Most believed this caused the sea level to drop 300 feet or more. Areas of the Bering Sea dried. Where there had been a sea, a land corridor stretched a thousand miles across that basin between the new and old worlds. Over those thousands of years of the predominance of ice in the north, people from the Siberian side likely ventured onto and then across that rolling land bridge following animals they hunted.

Some speculate the land in front of those people was an inhospitable place where in summer cold raging rivers gushed outward from the great Arctic ice sheet, and during winter icy winds swept the landscape.[13] However, no doubt life was not easy in Siberia either. Its tough human inhabitants were accustomed to it, and on they journeyed. Likely those hearty people didn't think of it as a new land–just land over the next hill.

The Upper Copper River country and Eastern Alaska Range where one now finds the villages of Chistochina, Slana, Twin Lakes, Mentasta, and the roadhouses and lodges along Nabesna Road was back then a land covered by hundreds of feet of ice. As time passed and the great glaciers melted, the Upper Copper River country was at least partly ice free. The people who had come to the coastal areas of the New World could venture inland.

These people who ventured from Asia were surely not that different from us, today. They, just as we, had their pioneers and adventurers. Some of those early people whose ancestors had crossed the Bering land bridge, led by their pioneers, likely journeyed into the upper Copper River country and settled where fish and game were plentiful. That was the beginning of Balzulnetas and Suslota Villages. The decendents of those early people with their rich histories are there in the Nabesna country today and known by the last names Justin, Nicolai, DeWitt, Joe, Ewan, Johns, Johnson, Bell, Hancock, Charley and others.

The Batzulnetas people certainly hunted the white sheep, the Dall sheep.[14] Those spectacular animals continue to seek out the high cliffs at Nabesna. Today, still, even with a flat-shooting rifle, hunting the white sheep is a difficult hunt because Dall sheep with their keen eyes can spot a predator miles away. Back before Europeans arrived with their guns, such a hunt was vastly more difficult, and wherever human cunning could substitute for the almost impossible stealth required to get near one of these animals, no doubt it did.

White Mountain, only 30 miles east of Batzulnetas offered a special advantage to early Nabesna hunters of the white sheep. Old timers say the Batzulnetas people knew those cliffs at Nabesna.[15] Some lance points and arrowheads of those ancient people were, according to locals, fashioned from flinty quartz that, like obsidian, breaks with razor sharp edges. This is the kind of quartz that formed in the post-volcanic stage at White Mountain and is found above those white cliffs. This coincidence leads one to believe that during some long ago hunt, ancient hunters made two important finds very near the same place up there on White Mountain.

Located up on that mountain, near the highest mine workings, is a cave called the "Sheep Cave". Sheep manure is on its floor many feet thick, likely having built-up over centuries or millennia. Sheep shelter in that cave today. In that cave the sharp eyes of the sheep were of little use. Then a stealthy hunter with bow and lance had an unusual advantage. Likely, while taking this advantage the hunters also found flinty quartz on a nearby windy ridge. That flinty quartz breaks with razor sharp edges. Some had yellow grains.

The cave and a nearby source of arrow and lance points may have had, long before the white man came, a special significance to the Batzulnetas Indians, in their hunts for meat and white sheep hides. The quartz still has a special kind of appeal. However, the early hunters did not dream of the meaning we now put on those yellow grains. The relative value systems of the early hunters who made arrow points from the quartz and today's people who see only the yellow in the same quartz is grist for debate.

---

[11] Interior and northern Alaska escaped continental glaciation.
[12] See, Hopkins, David M. (1973) "Sea level history in Beringia during the last 250,000 years," *Quaternary Research,* vol. 3. Fagan, Brian M. (1987) *The Great Journey, the peopling of ancient America,* Thames & Hudson, NY. Scott A. Elias (1995) *Ice Age history of Alaskan National Parks,* Smithsonian Institution Press, Wash. D.C.
[13] Fagan, Brian M. (1987) *The Great Journey,* Thames & Hudson, NY. The author, Professor of Archaeology at University of California at Santa Barbara writes of the peopling of Ancient America.
[14] The white Dall sheep was named for Dr. William H. Dall, a member of Robert Kennicott's party in 1865 organized by the Western Union Telegraph Co. to survey the route for a telegraph line from the U.S. across Canada and Russian-America (Alaska) to Russia.
[15] Carl F. Whitham, a long time friend of the Batzulnetas Indians wrote in the November 1, 1938, Report to shareholders of Nabesna Mining Corporation, "Long before the coming of the white man Indians knew these white cliffs as 'El Se Ba'." See Meyer, Mark P., et. al, (2000), *Mineral Assessment of Ahtna, Inc. Selections in Wrangell-St. Elias Nat. Park & Preserve.*

A natural cave called the "Sheep Cave" is in altered limestone near the upper mine portals. Dall sheep have sheltered in it for centuries. The Batzulnetas Indians, long before the coming of the white man may have found the cave and used it during their hunts for the prized white sheep. The cave may have played a role in the discovery of the gold-rich Bear Vein. The brown rock is limestone altered to garnets of various colors and the white rock is wollastonite. Both rock types at Nabesna are called "skarn."

CHAPTER 6

First Europeans Enter Nabesna Area

For thousands of years before any European knew Nabesna existed, the early peoples and their descendents inhabited the Nabesna. Sadly, these early people did not make permanent records of their activities and discoveries. What these early people might have told the late coming pale skinned people about their Nabesna home can only be dreamed of. Their collective knowledge of the natural history of this magnificent and mysterious country will almost certainly, unfortunately, never be matched by that of the late arriving Europeans.

One only can work with what one has. As a result one can know more about the most recent 150 years than about the previous several thousands of years. For it has only been within about 150 years that people made and kept records of their knowledge of the Nabesna.

We know that in 1845 Captain Michael Tebenkof of the Imperial Russian Navy, like one of his predecessors, Admiral Baron von Wrangell, was concerned by reports the English Hudson Bay Company had pushed into the eastern Alaska country claimed by Russia. That quasi-military English fur trading company had established Fort Selkirk in what is now Yukon Territory, Canada. The Russians believed the English were planning further moves to the north and into Russian territory.

To reconnoiter that distant country and determine what the English were up to, Governor Tebenkof,[16] himself an accomplished explorer, decided to send an exploration party in 1848 to the upper reaches of the Copper River country and beyond. He sent Lt. Ruffus Serbrinkoff, a creole, of the Russian Navy, with a small detachment (some say 11 soldiers) to carry out the reconnaissance.

Lt. Serbrinkoff ascended the Copper River. Thereafter, he was swallowed-up in the vast unexplored upper Copper River-Nabesna country.

Failing receiving word from or about Serbrinkof, the Russians considered him dead. They never again attempted to penetrate that part of Interior Alaska.

Following the purchase of Alaska by the United States from Russia in 1867, the United States made efforts to explore the new territory and to determine its resources and the amiability, or hostility, of its native peoples. General Nelson B. Miles, for whom Miles City, Montana, is named, was in charge of the U.S. Army in the Northwest. He was stationed at Sitka and had made an inspection tour of southeast Alaska coastal areas. In 1883, Miles ordered Lt. Frederick Schwatka[17] to carry out an exploration down the Yukon River through Alaska to the Bering Sea. That journey would become popularized by his book *Along Alaska's Great Rivers* and *Summer in Alaska*. White men were getting nearer the Nabesna area.

In 1884, General Miles ordered Lt. William Abercrombie to ascend the Copper River and explore routes through the Eastern Alaska Range. Lt. Abercrombie failed in this expedition.[18]

Gen. Miles next ordered his new aide, Lieutenant Henry T. Allen,[19] a recent West Point graduate, to proceed with the exploration of the Interior, accompanied by a sergeant and a private. Lt. Allen and his party landed at the mouth of the Copper River in March 1885. He traveled up the Copper River and eventually reached the confluence with the Slana River. It was early June 1885 when they got that far. They were then about 38 statute miles from what would become the Nabesna Mine.

Due to the effort of trekking that far Allen's party was weak and out of food. Local Indians, the Batzulnetas, found the party and guided it to Batzulnetas Village on the banks of Tanada Creek, a branch of the upper Copper River. Even now there is a trail to the location of that abandoned village and a sign on the Nabesna Road indicating the trail's departure point from the road. This village was a haven and perhaps the saving of the expedition.

At the village, Lt. Allen met Chief Batzulnetas. Allen described the chief as a warrior, broad shouldered, over 6 and a half feet tall, who wore his hair down to his waist. The chief was popular with his people. He wore a scarlet coat; one Lt. Allen believed to be of Hudson Bay origin.

Spring arrives late, but in a mighty rush, in the Nabesna. It was spring, after a hard winter. The villagers were also short of food. Nevertheless, they shared what they had with Lt. Allen and his men. The lieutenant and his men stayed until late June, the beginning of a time of plenty, when the Copper River was teeming with salmon just as it is every year to today. Little did they know, or care, they were dining on what today's gourmets call "Copper River reds."

Lt. Allen seemed more impressed with the Batzulnetas people than with

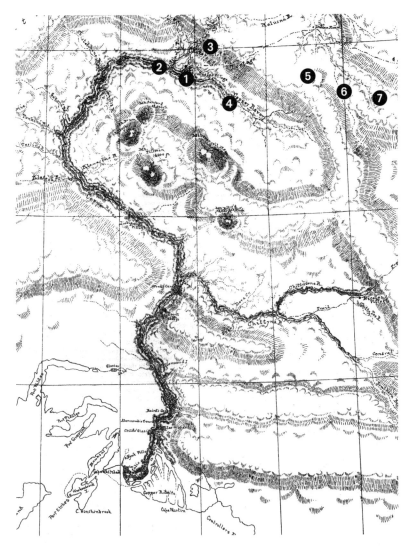

In 1885 Lt. Henry T. Allen, U.S. Army, and two soldiers made their historic reconnaissance through Alaska from the Gulf of Alaska to the Bering Sea. The Batzulnetas Indians rescued the starving explorers near the confluence of the Slana and the Copper River. Far different was the first encounter of Europeans when in 1848 Russian Lt. Serbrinkoff and men came to a very bad end at the hands of the Batzulnetas. Batzulnetas Indians guided Lt. Allen to White Mountain and what is now Nabesna Gold Mine suggesting the white cliffs or something else there may have held some special meaning for them. (From: *An Expedition to the Copper, Tanana and Koyukuk Rivers in the Territory of Alaska in the year 1885*, Washington, D.C., 1887).
1. Batzulnetas Village. 2. Junction of Slana & Copper River. 3. Suslota Pass. 4. Copper River. 5. Hatchers indicate cliffs of White Mountain at present Nabesna Gold Mine. 6. Jacksina Creek. 7. Nabesna River.

other Indians he met on his historic journey. The Batzulnetas, Lt. Allen reported, shunned the cheap Russian trade muskets used by Indians along the lower Copper River and instead fashioned their lance, arrow shafts, and bows from fire-hardened willow. He described them as warlike people who controlled the passes through the Alaska Range and extracted tribute from other Indians using the routes between the Copper and Tanana Rivers.

They also differed from the other Indians in another respect–one that Allen found singularly intriguing. The Batzulnetas did not speak the language spoken by those peoples of the lower Copper River. Lt. Allen wrote in his War Department Journal the words and phrases used by the Batzulnetas people were closer to Apache[20] spoken in the southwestern U.S. territories than to other native languages he found in Alaska. His most startling discoveries, however, came when Chief Batzulnetas gave him Lt. Serbrinkof's journal (1848).

For the first time the Batzulnetas told outsiders, and that included other Indians in the Copper River area, that long ago strangers had come to their village and demanded food and women. The Batzulnetas were not to be intimidated. The warriors fought and killed the arrogant Russians. These Russians, who perhaps had become accustomed to having their way with indigenous peoples farther to the south, thought they would find the same docility in these upper River people, greatly underestimating the Batzulnetas.

The last notation in the Russian's journal was "62 degrees 48 minutes and 45 seconds." That location is a few miles from Batzulnetas and Nabesna Mine. The fate of the lost Russian exploration was known.[21]

While at Batzulnetas, guides must have taken Lt. Allen to what is now Nabesna Mine. On his 1885 map, Lt. Allen drew hatches representing a mountain that juts out and overlooks the junction of two rivers, as does White Mountain. Those rivers can only be Jacksina Creek and Nabesna River, which are plainly visible from the old mining town at the base of White Mountain. The mountain depicted by Lt. Allen can reasonably only be White Mountain. The Batzulnetas could have guided the white men to many places, but they choose to take them to White Mountain. This supports the speculation that the cliffs and sheep cave with their hunting opportunity and a source of arrow and lance points had a special meaning to the Batzulnetas people.

After the arrival of the salmon in the upper Copper River on their annual spawning run up the river in 1885, the Batzulnetas and Lt. Allen and his men had food. The soldiers were fit to travel. The Batzulnetas guides led Lt. Allen's party to Suslota Lake Pass. From there, the guides took them past Mentasta Lake, then through the Alaska Range and finally down the Tok

River to the Tanana River.

At the Tanana Lt. Allen and his men bid farewell to their benefactors. There, he and his men built a raft and floated on it down the Tanana to the Yukon River, then down the Yukon. After several side expeditions they floated to St. Michael on the Bering Sea. There they caught a steamer for Seattle. Nabesna was on the map.

Visitors now traveling Nabesna Road may stop at Mile Post 9 where in 1885 Lt. Allen and his men crossed what is now Nabesna Road on their 15 mile walk from Batzulnetas Village across that tundra and brushy plain to Suslota Pass in the Alaska Range. In all the intervening time little has changed. The present visitor will see the same sights, as did Lt. Allen and his men when they started their historic crossing of the Eastern Alaska Range.

Lt. Allen put his personal stamp on the map of the area. West of Nabesna Road rise the great snow capped Wrangell Mountains. The nearest, and most spectacularly visible, is 16,208 foot Mt. Sanford, named by Lt. Allen for his storekeeper grandfather.[22, 23] Lt. Allen also named Miles Glacier on the lower Copper River for General Miles and in the Wrangells named mountains for other friends.

For twelve years after Lt. Allen left their village, the Batzulnetas pretty much had the Nabesna to themselves, and they might still but for an event that started in motion an influx of people that forever changed that north country.

---

[16] Most governors of Russian-America after Alexander Baranov were officers in the Russian Navy and served five year terms. Navy Captain Michael Tebenkof was appointed governor in 1845. He was a noted explorer of the coastline of Alaska. Under his direction was published the first Hydrographic Atlas that included 38 charts all engraved on copper plates by Tarenlief, an Aleut and published in 1852. Brooks, A.H. (1953) *Blazing Alaska Trails,* Univ. of Alaska, and Maturin M. Ballou (1898) *The New Eldorado*: Houghton Mifflin & Co. Boston.

[17] Lt. Frederick Schwatka was only a baby when with five sisters his parents crossed the Great Plains by covered wagons to Oregon in 1853. He graduated from West Point in 1871 and served throughout the west. His greatest fame was his explorations in Alaska and the Arctic. His death came unexpectedly in Portland, Oregon when he was only 43. The Schwatka Mountains are named for him.

[18] Captain Abercrombie was in 1899 ordered back to Alaska to survey a military route from Valdez to the Yukon River in part to aid prospectors during the gold rush. Captain Abercrombie would with USGS surveyor Oscar Rohn play a role (some say the important role) in the 1900 discovery of the copper deposits on which the famous Kennecott Mines were built. Abercrombie Mountain near Valdez is named for him.

[19] Lt. Allen was born in Kentucky in 1859 and graduated from West Point in 1882. After Alaska he rose to rank of Lt. General and served with distinction in the Philippines during and after the Spanish-American War. He died in 1930. Mt. Allen, a few miles east of Nabesna Gold Mine, is named for him.

[20] Judge James Wickersham, a U.S. District judge in the early days of the Territory of Alaska, was also a recorder of history. He ascribed to Father Jette's; S.J. writing Athabascan people migrated from Alaska to what is now Arizona and beyond. Father Jette believed the eastern Alaska Indians were blood brothers and sisters and spoke the same stock language as did the Apache. If that were so, what Whitham said to be the old Indian name, or the anglicized version of "El Se Ba," for the cliffs at Nabesna, and what others say can only be Apache, is not necessarily a mystery. Father Julius Jette, S J. served as a long ago missionary in eastern Alaska.

[21] Pierce, R.A. and Alton S. Donnelly, (1978), *A History of the Russian-American Company*. University of Washington Press, Seattle, p. 352. A translated and edited the work of Tikhmenev's 1861-63 history of the Russian-American Company, in which the authors describe Serbrinnkoff's (Pierce's spelling) death, indicating it was known before Lt. Allen reported it.

[22] Lt. Allen was a prolific namer of places in Alaska. In the Wrangell Mountains he named Mt. Wrangell (already named Mount Wrangell by the Russians for Baron von Wrangell, one-time governor of Russian-America), Mount Tillman for professor Samuel Escue Tillman of West Point, Mount Blackburn for U.S. Senator Joseph Clay Stiles Blackburn from Kentucky (Lt. Allen was from Kentucky), and Mount Drum for General Richard Coulter Drum of Civil War fame, and many others throughout Alaska.

[23] Judge Wickersham in 1938 wrote the Indian name for Mt. Sanford was "Kuhltan." Wickersham, Hon. James, (1938), *OLD YUKON Tales-Trails-and Trials*: Washington Law Book Co., Washington, D.C. p.446

A bit of mystery is added to the Serebenikof (USGS spelling) saga by Brooks when he writes (page 236, *Blazing Alaska Trails)*, that the Hudson Bay Co. reported natives told that Russians had reached the Yukon in 1848. They did so following a northeast flowing river. Brooks believes it could only have been Serebenikof and the Fortymile River. The Fortymile River area and the adjoining Klondike district in Canada would a half century later become one the richest gold placer areas in the World. Brooks believes Serebenikof was on his way back when killed by the Batzulnetas. If Serebenikof had discovered placer gold on the rich Fortymile and had been able to report it to Governor Tebenkof and the richness had become as widely known as the then 1849 California gold rush, one might wonder if Russia would have 19 years later sold Alaska to the United States?

CHAPTER 7

THE GOLD RUSH ERA
Early Prospectors Enter Nabesna Area

In 1897, the steamer S.S. Portland from Alaska docked at Seattle. Word spread that tons of gold was aboard. A northern gold rush was born. The Klondike.

By the thousands men and women from all walks of life, states and some foreign countries made their way north to reach the Klondike gold fields and build a rich new life. Most went by way of Skagway, Alaska, and over White Pass or Chilkoot Pass to enter Yukon Territory, Canada, and then down the Yukon River to the Klondike gold fields near Dawson. Some even went by way of stern wheelers up the Yukon River.

Not all Klondikers chose Skagway or the Yukon as their routes to the gold fields. Hundreds traveled by steamer to Valdez[24] as their staging point for their trips to Dawson in the Yukon. From Valdez the remaining 450-mile journey to the Klondike was no less hazardous than the better-known, more southern, journey from Skagway. It may have been more so, accounting for the greater popularity of the Chilkoot route. Crossing Valdez Glacier just as they started their trek was only the first of many perils.

Nevertheless, those early Klondikers opened that part of Alaska, found the copper deposits that became Kennecott Mines and discovered other mineral deposits. The trails they blazed became Alaska's important highways including the Richardson, Edgerton, Glenn, Slana-Tok Cutoff, Taylor and, not the least of which, to anyone who has ever traveled it, the 46-mile long Nabesna Road.

Many of these impetuous adventurers were unprepared financially and by training and experience for the task they had set for themselves. Poorly equipped, many were soon broke. To worsen matters the winter of '98-'99

Remains of one of the old "White Mountain Cabins" built by the first prospectors in 1898 at what is now Nabesna Gold Mine. The cabins are some of the oldest in eastern Alaska. Carl F. Whitham used the cabins in the twenties. This photo was taken in 2000. The young man is the grandson of the present owner of Nabesna Gold Mine.

was severe. The result was that hundreds of gold seekers found nothing but grief. Many became broken, destitute and scurvy-ridden. In that year the U.S. Army built a hospital at Valdez, provided soup kitchens and arranged to transport men back to Seattle.

Many others, those better equipped, those luckier, or just tougher, pulled through those hard times and stayed on in the new country. They found placer gold at the head of the Chistochina River[25] and in the Eastern Alaska Range. They found other prospects in that wild country between Valdez and the Klondike. This initial success encouraged later arriving men to find their ways into that vast interior intending to mine and then later, when they could not imagine leaving this entrancing country, to stay.

One party out of Valdez seeking gold in that wild country found its way to White Mountain and what is now Nabesna. Its members found placer gold in two gulches that they later called Discovery and Stamp Mill Gulch. The presence of the placer gold led to the finding of a 10-foot wide gold-bearing quartz vein above Stamp Mill Gulch on the lower slope of White Mountain. Years later the vein would be called "Tower Knob Extension."

Those early prospectors in 1898 built the first log cabins, probably the

The stamp mill near the White Mountain Cabins was the first of its kind in eastern Alaska and was freighted there by dogs and horses from the coast at the turn of the century by The Royal Development Corporation. Ore mined and milled graded two to three ounces gold per ton but USGS (1907) reported stamp mill recovery was poor and the effort abandoned.

first permanent structures in the Nabesna. Considering their wide notice, those cabins very likely were the first log cabins in that part of Alaska. One of those old cabins of bleached logs with its collapsed roof stands in the brush a half-mile from the old mining town of Nabesna. Evidence of others is nearby. So few were cabins in that part of eastern Alaska these collectively became a landmark known as the "White Mountain Cabins."

In 1903, the US Army built a trail from Valdez to Eagle on the Yukon River in far eastern Alaska Territory. The Army called it the Eagle Trail.[26] That made travel easier than previously to the interior, as well as to White Mountain. A prospector heading to White Mountain left the Eagle Trail at Ahtell Creek at what is now the community of Slana and traveled southeasterly towards Nabesna River.

Several of those early prospectors at White Mountain organized the Royal Development Corporation in Seattle and sold shares. In the winter after organizing and accumulating capital for their venture, those men used the EagleTrail to freight in from Valdez a two-ton stamp mill purchased with their venture capital. They set up the mill on Stamp Mill Gulch and continued prospecting. They drove several short drifts, called adits, into the mountain. One 30-foot deep adit may have played a part in a bizarre clandestine act almost a half century later. On the steep slope above Stamp Mill Gulch, these early miners opened a 10-foot wide gold-bearing quartz vein by trenching. They milled the gold ore in the stamp mill. The mill still stands along side Stamp Mill Gulch. The vein from which these men mined their ore is visible on the mountain slope and gold is still found there.

About this time came a rider on horseback leading a packhorse along the same Eagle Trail. For the last few days snow-covered Mt. Sanford across the Copper River was on his right and on his left were the rolling foothills of the Eastern Alaska Range. He'd not been here before, but knew the country through tales of prospectors and works of the U.S. Geological Survey.

The year was 1907. The rider's name was Carl F. Whitham. For seven years since his arrival in Valdez, at age 18, little is known of Whitham's activities or of his exact whereabouts. However, he had come to Alaska in search of gold, and it is virtually certain in that time he was involved in mining, probably in the Valdez area. But regardless of his inauspicious arrival, in the years to come, Carl F. Whitham would leave a greater legacy in the Slana-Nabesna area than has any other person since his arrival there.

He turned off the trail into Ahtell Creek valley to prospect the silver-lead deposits he'd heard tell of. Later, he crossed the Slana River and came to know the Batzulnetas people.[27] Thus he began a life-long friendship with them. He prospected the mountains, called Mentasta Mountains[28] within the Eastern Alaska Range, a few miles east of Batzulnetas Village and the present Nabesna Road. There Whitham staked a number of mining claims, including what he called the Caribou-Comstock covering gold bearing veins. He drove several short drifts on the claims and to pay expenses sluiced gold from at least one creek in the area.

Whitham, probably about this time, met D.C. 'Bud' Sargent. Sargent was one of the early prospectors in that country. He entered eastern Alaska in the early 1890s. Sargent discovered the copper deposit called Orange Hill[29] at the terminus of Nabesna Glacier, some 12 miles south of present day Nabesna. Some old maps show what is now the town of Nabesna, or what then might have been the "White Mountain Cabins," was at the turn of the century called "Sargent's Camp". The history of U.S. Post Offices in Alaska describes D.C. Sargent as the postmaster at "Nabesna" in 1909.[30] In the coming years Whitham and Sargent became good friends.

For several years Whitham prospected and learned the Nabesna country. In that process in the spring of 1912, Whitham made arrangements in Chitina and paid the General Land Office to survey his Caribou-Comstock

D.C. 'Bud' Sargent, left, came to Nabesna country before the turn of the century. He was one of the locators of Orange Hill copper deposit. Town of Nabesna was at turn of the century called "Sargent's Camp." Whitham probably met Sargent in the Nabesna area about 1907. Photo was taken by Whitham at Nabesna 1929.

claims leading to patent.

That he did presents an interesting conundrum. Mineral surveying was then, as it is now, an expensive and difficult process. Entering upon that process indicated an abiding intention to patent in the person who did so. The Land Office, at Whitham's request, surveyed the claims in 1912 and assigned it U.S. Mineral Survey No. 972. Nevertheless, records fail to show Whitham ever applied for a patent.

One can infer in these and later measures taken by Whitham that he was discouraged or at least needed replenishment. Whitham in the fall of 1912 traveled back along the Eagle Trail to the junction of the Richardson Road and then followed it until he reached the road to Chitina. From there he followed the trail along the railroad to McCarthy and proceeded on to the Mother Lode Copper Mine up McCarthy Creek. At the mine he hired on as a miner. The Mother Lode was on the opposite side of the ridge from the Bonanza and Jumbo Mines that later became owned by Kennecott Copper Corp.[31] James E. Godfrey was president of the Mother Lode Copper Mine and Whitham met him for the first time. Godfrey would years later play a role in Whitham's affairs–but not a pleasant one.

In the meantime, in 1909, the USGS reports visiting the Royal Development prospect at White Mountain. No one was there.[32] According to the USGS about 60 tons of ore was mined from the vein and run through the stamp mill. USGS estimated the ore to grade one to three ounces of gold per ton, but the stamp mill operation was said to be inefficient and loss of gold was high. The USGS records that The Royal Development Corporation continued assessment work on the mining claims to 1914. Then it allowed the claims to lapse. According to the USGS, various prospectors in the following years relocated some of the claims and then dropped them.[33]

---

[24] First called "Copper City," then in1899 the name was changed to Valdez.
[25] See Elizabeth A. Tower, 1996, *Ice Bound Empire,* Anchorage, Alaska .
[26] Called the "Eagle Military and Telegraph Trail," it was staked under Capt. Abercrombie's command in 1903 between Valdez and Fort Egbert at Eagle City on the Yukon River. Near Slana, the Glenn Highway (locally called the Slana-Tok Cutoff) follows close to sections of the old Trail. At Ahtell Creek the trail turns north and follows Porcupine Creek to Indian Pass and crosses the Slana River Valley upstream from Mentasta Lake to continue over Sikonsina Pass, and then it crosses the Tok River and follows it to the Tanana River. Stretches of the old trail are still to be seen and have been deepened over the many decades by animals. In the Tok River country old pole tripods still hold strands of telegraph/telephone wire.
[27] National Park Service in 2000 briefly discusses historic Native settlements in the Slana-Nabesna area but never mentions the village of Batzulnetas, which likely was the most important Native village in all that area. See page 34, *Environmental Assessment, April 2000, Mining Plan of Operations for Nabesna Mine, Wrangell-St. Elias.*
[28] D.C. Witherspoon of USGS used the name in 1902. F.C. Schrader, USGS, believed it to be based on the Indian name, "Mantasna".
[29] Named by F.C. Schrader, USGS in 1902.
[30] See Ricks, Melvin B., (1965), *Directory of Alaska post office and postmasters,* 1865-1963; Ketchikan, Alaska.
[31] Rohm, USGS, in 1899 named Kennicott Glacier for Robert Kennicott, Director of Scientific Corps in 1865 exploring a route through Alaska, then Russian-America, for the Western Union Telegraph Expedition. Kennecott Copper Corp. years later took its name from Kennicott Glacier but misspelled it with a second "e." The town of Kennecott took its name from the mining company. Later the name was changed to that of the glacier and "i" replaced the second "e." USGS Prof. Paper 567.
[32] Moffit, F.H., et al. 1910, *Mineral Resources of the Nabesna-White River district, Alaska*: USGS Bull. 417.
[33] Moffit, F.H., 1943, *Geology of the Nutzotin Mountains, Alaska and Gold Deposits Near Nabesna.* USGS, Bull 933-B.

CHAPTER 8

The Shushana Gold Rush: A Prelude to Nabesna Mine

According to M.J. Kirchhoff,[34] three miners walked 60 miles over the mountains from Shushana, later called Chisana, and arrived in McCarthy in July 1913. This was less than a year since Whitham came south from Nabesna to the copper mine to work for wages. These men told of the rich placer finds at Shushana. To bolster what they claimed, they bought supplies with gold. Word of gold quickly spread through the Kennecott Mines and the Mother Lode mine where Whitham was working.

The rush was on. Several thousand stampeders struck out for the new fields. Whitham, too, heard the news of gold in Shushana. If Whitham's intention was not to return to the Nabesna, this news seemed to have urged him toward his destiny

Shushana was on the north side of the Wrangell Mountains from the copper mines. Prospectors could reach the new fields by a trail across the Wrangells so steep in places as to be called the "goat trail." However, there was another route. It took the traveler around the mountains, past Batzulnetas, past White Mountain, across Nabesna River and then the remaining 40 miles to Shushana. While in truth this route was much longer than the "goat trail," on this route through the Nabesna one could with saddle and packhorses carry supplies needed for mining. Likely, equally important to Whitham, was the fact that he knew this trail. Whitham joined in the rush but took the longer route. He never again throughout his days truly left the Nabesna area once there this time.

He followed the old Eagle Trail to Ahtell Creek where he left the trail and forded the Slana River where the present bridge is today. He continued along the trail to Batzulnetas Village. After visiting with his old friends he journeyed the 40 miles further on to the Nabesna River. Years later

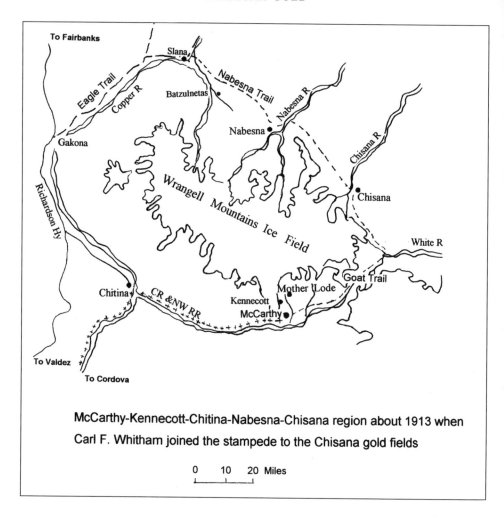

McCarthy-Kennecott-Chitina-Nabesna-Chisana region about 1913 when Carl F. Whitham joined the stampede to the Chisana gold fields

Whitham followed parts of that same trail as he blazed the route for Nabesna Road from the Slana River to Nabesna Mine. After fording Nabesna River he continued another 40 miles to the new gold fields.

At Shushana he staked his first placer claim on Eldorado Creek, and later acquired claims on Skookum and Snow gulches. Unlike many of the other gold seekers, he packed with him equipment and supplies he needed. He was ready to mine. Whitham was in his early thirties, a hard worker and had mining experience. Working his own mine was why he had come to Alaska.

These gold stampedes certainly did not uniformly result in happy endings for all their participants. As so often happened in other gold stampedes, those first describing the gold exaggerated the richness of the creeks. Exaggerated richness translated to unduly large numbers of gold seekers. The result was that only a few men made any money. George Hazelet,[35] a

Photo by Carl F. Whitham of his Chisana placer mine in 1920. In 1913, Whitham left the Mother Lode Copper Mines of Alaska near Kennecott for the gold rush to Shushana, now called Chisana. Chisana is on the north side of the Wrangell Mountains and about 40 miles east of Nabesna.

prominent citizen of early day Valdez and Cordova, visited the Chisana country and later mentioned Whitham's placer claim as one of the better ones. Whitham was one of the few who actually made money from this rush.

Whitham built his cabin on his claim. When winter came and the creeks froze he sometimes stayed and trapped. Other times he rode horseback back to Chitina where he could take the Copper River and Northwestern Railroad from there to Cordova and the steamship to Seattle, where he sometimes wintered. Kennecott Company eventually controlled both railroad and steamship line. Whitham during those years when he traveled "outside" returned to his claims in the spring just before ice broke up.

Whitham explored his ground in a miner-like manner. He dug prospect holes and then panned the gravel and recorded in his ledger the value of gold in the gravel recovered from each hole. His prospecting showed him where the paystreak was. A number of men worked for him, some freighting supplies, some working at the mine, others whipsawing lumber and still others cutting and hauling firewood. His ledger of 1915 shows he paid them in gold dust and used gold to pay for supplies. Many of the men who worked for him at Chisana would work for him again years later.

He met and became friends with a number of successful placer mine owners at Chisana. Such a mine owner was D.P. Thornton from Hot

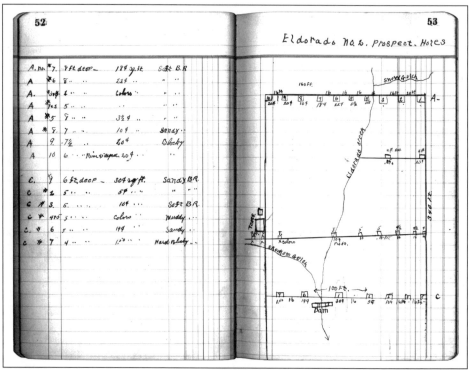

Whitham's 1915 ledger shows he dug prospect holes, panned the gravel from each hole, and determined where the gold paystreak was. He then mined that paystreak. His claims were reported to be some of the better ones at Shushana.

Springs, Arkansas. Thornton had a daughter who remained in Arkansas perhaps dreaming of the romantic place where her father had gone. One has to believe that Thornton wrote his daughter of his friends, including Whitham, in letters home.

Whitham knew of the early prospectors at White Mountain and the Royal Development work there. During those years when he was returning to Chisana he stopped at White Mountain. While at the mountain during his Chisana years he looked at the old prospects, and collected samples of gold bearing quartz. He crushed and panned the samples and by weighing the gold on a pair of scales he could estimate what the rock ran in ounces of gold per ton. His gold scales, like those carried by many prospectors, were small and could be folded-up and carried in a pocket. It was during these stops that Whitham must have determined that White Mountain held promise and would play a part in his life.

In 1915, World War I erupted in Europe. The United States was soon drawn in. Whitham then leased his placer claims to a friend, took the steamship to Seattle and enlisted in the U.S. Army.

When the war ended he was honorably discharged on the East Coast. He had in the bank $7,000 earnings from his placer claims. After visiting a brother, as well as James Godfrey of the Mother Lode Copper Mine, in New York City, he took the train to Hot Springs, Arkansas. There, in March 1919, he and Marie L. Thornton were married. One wonders if Marie buoyed Whitham with letters while he soldiered in Europe.

Whitham and Marie journeyed to Seattle. In August they boarded the Alaska Steamship "Yukon" for Alaska. Their first class ticket to Cordova cost $86. The cruise took them through the scenic inland passage of southeast Alaska. It was Marie Whitham's first ocean cruise. She spoke glowingly of it.

After stops at several coastal towns including Ketchikan and Juneau they docked at Cordova. There they stayed in the Windsor Hotel and after a week of buying supplies they boarded the Copper River and Northwestern Railroad and rode the 90 miles to Chitina. There the newlyweds spent a week in Chitina where Whitham visited old friends and introduced his bride. After purchasing more supplies Whitham and Marie rode the automobile stage that traveled up the Edgerton Highway (named for General Edgerton) to the junction with the Richardson Highway. The Richardson Highway linked Valdez with Fairbanks. However, they went only as far as Gakona Roadhouse.[36] Whitham had telegraphed from Cordova and agreed to purchase three packhorses and two saddle horses. There the journey took on a more arduous quality.

Riding saddle horses and leading the packhorses they started their 175-mile long journey from Gakona to Chisana. They followed the old Eagle Trail and had to ford the wide Chistochina River and finally reached the Slana River. Years later the Glenn Highway, the Slana-Tok Cutoff, would follow much the same route. They forded the Slana River but the widest and most dangerous was the Nabesna River 50 miles further on. "The bear came right out of the brush and knocked one of the packhorses down", Marie wrote a friend, "Carl shot it." This was their honeymoon. She wrote how they camped along the trail, about the long horseback ride, of the rain and snow at the high places.

In September 1919 they reached Whitham's gold placer claim and their long journey ended. Whitham took back his mining claims. He and Marie mined until 1922, when he again leased out his claims. Whitham apparently felt his destiny was back down the trail. He and Marie left Chisana[37] and moved to White Mountain where she setup housekeeping in one of the old White Mountain Cabins along Cabin Creek.

Whitham and Marie spent the winter of 1922-1923 in Chitina. Returning in the spring to White Mountain they found a grizzly had torn open the roof and ruined the inside of one of the cabins.

That year, 1922, Whitham staked his first lode claims at Nabesna and called them "White Mountain" claims. People know them by that today. In the fall they left White Mountain and rode horseback to Chitina where they rented a small house and spent the winter of 1922-1923. In the spring they returned to White Mountain. On arrival they discovered a grizzly bear had torn through the roof of one cabin and destroyed the inside. During the summer Whitham prospected the old workings. Marie never complained, but Whitham knew this was no life for a lady.

In August 1923, they left White Mountain for Chitina the way they had arrived, on two saddle horses leading two packhorses. They camped several days near his 1912 Caribou-Comstock claims. There he collected more samples. They visited with his friends at Batzulnetas before going on to the confluence of Slana and Copper Rivers. They camped for a few days in the upper Ahtell Creek area. He again prospected there as he had years ago when he came to the area in 1907.

Leaving the Ahtell-Slana area they rode the last 70 miles to Gakona Roadhouse. They left the horses there and took the automobile stage to Chitina. From there they rode the railroad to Cordova and then boarded the Steamship to Seattle.

In Seattle, Carl and Marie eventually purchased a house in nearby Renton. Before returning to Alaska in the spring of 1924, Whitham had

A.L. Glover, Inc. assay the last of his samples collected at the Caribou-Comstock claims on their way out the previous fall. The Certificate of Assay dated April 24, 1924, showed "Sample #2. Caribou-Comstock," assayed 42.60 oz. gold per ton and silver 33.40 oz.[38]

---

[34] Kirchhoff. M.J. 1993, *Historic McCarthy*, Juneau, Alaska.
[35] George Cheever Hazelet's life in early day Alaska is described in Elizabeth A. Tower's 1996 book *Icebound Empire*, Anchorage, Alaska.
[36] Gakona Roadhouse is one of the oldest on the present Glenn Highway. In February 1905 Judge Wickersham with a dog team on his way from Valdez to hold court in Fairbanks stopped here. He called it a good roadhouse. Many prominent Alaskans have over the many decades stayed there. It is a historic roadhouse and is listed on the National Register of Historic Places.
[37] Chisana is presently an attractive fly-in destination for the adventures. Little has changed since the gold rush.
[38] Sample #2 is exceptionally rich. Presently the ore would have a value of over $11,000 per ton. Regarding the Caribou-Comstock claims the question remains: why after spending a large sum of money in 1912 to have the claims surveyed for patent (the most expensive part of the process) did he never apply for patent? Then 12 years later comes this rich sample but yet there is no record he ever prospected further the claims. The site of the claims could have had something to do with the Batzulnetas Indians. They were his friends. Some believe the sample was collected near but not on the Caribou claim. That possibility is strengthened by BLM Technical Report 34, May 2000 that describes when in 1999 geologist sampled what they believed was the "Caribou" claim site but found no significant gold values.

A. L. GLOVER, INC.

SUCCESSOR TO GLOVER, WELLS & ELMENDORF, INC.
ESTABLISHED 1916

ANALYSTS—ASSAYERS—METALLURGISTS
615 PREFONTAINE BUILDING
SEATTLE, WASHINGTON, U.S.A.

CERTIFICATE OF ASSAY

No. 9567-68   Date, April 28th, 1924.

The Sample of Ore.
From   C.F. Whitham, 4002 California Ave., Seattle, Wash.
Marked   as below:-

and submitted to us for analysis, contains:

Sample #1. Toby Claim.

    Gold -------- 1.29 oz. per ton ---- Value $ 25.80
    Silver ------- 1.80 " " " ---- " 1.15
                                                               Total " $ 26.95

Sample #2. Caribou - Comstock.

    Gold -------- 42.60 oz. per ton ------ Value $ 852.00
    Silver ------- 33.40 " " " " 21.38
    Lead -------------------- 50.5 % ---- " 90.90
                                                               Total " $ 964.28

Gold   @ $ 20.00
Silver   @   64¢
Lead   @   09¢

Respectfully Submitted,

A.L. GLOVER, Inc.,

Charges $ 6.00   By *[signature]*

In the fall of 1923, Whitham took Marie to Seattle where they purchased a house. On the way he collected more samples at his Caribou-Comstock claim near Batzulnetas. One sample assayed 42 ounces gold per ton. Today, the ore would have a value of $11,000 per ton. A mystery remains why he didn't follow through with those claims. Moreover, there is no evidence he ever returned to the Caribou-Comstock claims. His action may have had something to do with his friends, the Batzulnetas Indians.

CHAPTER 9

Roads, Schools and People

In May of 1924 Whitham returned to Alaska, but there is no record Marie ever did. One has to suppose she enjoyed her new home in relatively civilized Renton. Whitham again settled at White Mountain and continued prospecting the old workings. He had money in Cordova and Seattle banks and his leased-out Chisana placer claim always paid money. Even so, prospecting raises no money and to insure funds for his wife and to support his prospecting he used part of each season working elsewhere for wages.

Whitham seems to have been blessed with an unusual foresightedness. In addition to taking care of Marie's needs, Whitham was interested in the well being of the entire Nabesna area. This he manifested from a time well before he had a mine there. He seems never to have lost sight of his vision of eastern Alaska with a road system. He knew without roads that nobody would live there. Plainly, Whitham observed that elsewhere other roads were building. He wanted at least a fair share for Nabesna.

Captain Richardson, the first president of the Alaska Road Commission (ARC), in about 1906 started the sled road, later named The Richardson Highway, from Valdez to Fairbanks.[39] The Richardson Highway was a good road from the coast to Fairbanks. Whitham knew that if the trail from the Richardson Highway to Chisana gold fields could be improved the whole country would be helped,[40] including his claims at White Mountain.

Whitham's long time mining friend and delegate to Congress, Dan Sutherland, favored roads too.[41] So also did members of the Territorial Road Commission[42] and the federal Alaska Road Commission (ARC). The ARC's purpose was to build and maintain federal roads in Alaska for the military and help the civilians develop the territory.

Whitham's first concern was a road from the Eagle Trail to the Nabesna River. To get money from Congress to improve the Nabesna trail and build it into a road Whitham knew he faced another problem. That was the other Alaska gold fields like Fairbanks, Nome, Flat, Iditarod, and Circle. They, too, lobbied for roads and had more votes and clout with Delegate Sutherland and the Board of the Alaska Road Commission than he had.

Three Army officers headed the Alaska Road Commission (ARC), not to be confused with the Territorial Road Commission. In 1924, Colonel J.G. Steese succeeded Col. Richardson[43] as president of the ARC. When he did Colonel Steese became the focus of Whitham's persistent attention. The Steese Highway from Fairbanks to Circle is named for Col. Steese.

To get a road to pass by White Mountain on the way to Chisana, Whitham first needed a bridge across the Slana River. Whitham pressed Col. Steese for the ARC to build one. On December 30, 1924, Col. Steese wrote Whitham from his office in the Munitions Building, Washington, D.C.:[44]

> "Concerning the proposed Slana bridge . . . I have referred the matter to Major Oliver" . . . "I would suggest that on your way through Juneau in March you call at the office to see Major Oliver or myself, and can go over the matter in detail...."

Whitham did, thereupon, Col. Steese and Major Oliver requested funds for a log bridge[45] to cross the Slana and for funds to improve the trail to Nabesna River, now followed in part by Nabesna Road. On December 17, 1925, Major Oliver, Engineer Officer for the ARC wrote Whitham,

> "I am advising you at this time that it is definitely planned to allot $2,000.00 for the work on the summer and winter trails between the Slana and Nabesna Rivers. We would like to have you arrange for and inspect this work . . ."

Whitham saw that the work was done.

During this time Whitham was also trying to interest mining companies in what he called the "Slana-Nabesna" or the "North Wrangell District." If a large mining company started work on a prospect there, it would help with lobbying efforts for the road. He obtained letters of introduction to New York firms from Stephen Birch, President of Kennecott Copper Corp. and E. T. Standard, Vice President of Alaska Steamship Company and later president of Kennecott. In November, 1925, E.T. Standard wrote Professor Alan

M. Bateman[46] of Yale University who also served as a consulting geologist for Kennecott. The letter introduced Whitham saying," I have tentatively agreed with Mr. Whitham that you will make a trip with him in to his prospects while you are in the North next summer." The prospects included the lead-silver deposits in the Slana-Ahtell Creek area. The Seattle Chamber of Commerce and Seattle businessmen also helped with letters introducing Whitham to eastern Senators and New York bankers.

Professor Batemen did look at some deposits in the area, but not Nabesna. He recommended if a railroad[47] were pushed through that region to the Tanana Valley, Kennecott should consider exploration there. Stephen Birch had at the turn of the century traveled through that area and probably favored a railroad.[48]

In the winter months, Whitham returned to Seattle. During these times away from Nabesna he submitted articles of general interest about Alaska to *Colliers, Saturday Evening Post* and *Liberty* magazines, but all he got back from the general circulation magazines was "no thanks." The mining journals did publish his "newsy" articles describing prospects in eastern Alaska. Newspapers, including the *New York Times* published some of his articles. Marie typed all of them

Whitham knew he was out-classed by the other Alaska mining districts in competition for funds for what he was in the mid-1920s calling "Nabesna Road." He figured there was more ways to skin this cat than one. If a mail route were established from the Richardson Highway to Slana and past Nabesna to Chisana gold fields, the ARC and Congress would be influenced to appropriate funds to upgrade the trail to a road.

The existing mail route to Chisana was from McCarthy, near Kennecott, by the infamous "goat trail" over the Wrangell Mountains. Whitham knew it was far too steep and craggy to ever become a road. He persuaded his old mining friends in Chisana to support a new mail route from the Richardson Highway through Slana past Nabesna and across the Nabesna River to Chisana. They did, and in a January 9, 1925, edition of the *Alaska Weekly* the paper told how because of the difficult mountainous route from McCarthy over the "goat trail" to Chisana the mail, while supposed to be delivered every 30 days, was sometimes two or more months late due to winter storms.

Whitham bid for and got the new route. The post office agreed to pay a flat rate to the mail carrier. His friend, Lawrence DeWitt, who built the Slana Roadhouse, would use dog sled in the winter and packhorses in the summer to deliver the mail. After a flat rate was agreed upon the post office changed the rules and came up with a shocker. They would not pay a flat

Present photo of the historic Slana Roadhouse. It played a role in the development of Nabesna Mine. The 1938 USGS map shows how Nabesna Road continued past the roadhouse to Nabesna Mine. During World War II the Army started at Nabesna Road near the roadhouse and constructed a road through the Eastern Alaska Range to the newly built Alaska Highway. The lower 1960 USGS map shows how the present Glenn Highway runs past Slana.

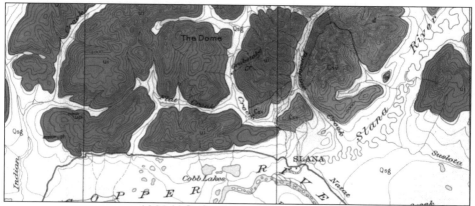

1938 U.S. Geological Survey Map

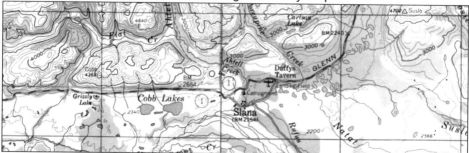

1960 U.S. Geological Survey Map

rate to the mail carrier but only pay per pound of mail carried! Plainly, per pound compensation would not support the effort required. In a "My dear Carl" letter of May 22, 1926, Delegate Dan Sutherland wrote:

> "You will recall that when you were here last spring they informed me at the Department (Post Office) that the bid submitted by a carrier would be accepted and this is the first intimation I have that their plans were changed."

Delegate Sutherland got the bid straightened out, but not before DeWitt expressed his feelings about government officials. Whitham enjoyed repeating:

> "Half of what they say is lies and the other half ain't true."
> (Some might say that's relevant today.)

Whitham did not spend all this time courting the road commissioners. He also helped establish schools where there were none in the Copper River country. To this end, while in New York, he sought help from the Board of Missions of Protestant and Catholic churches alike. He met with Miss Edna Voss, Secretary of the Presbyterian Church. She explained that according to

Lawrence DeWitt, owner of the Slana Roadhouse, carried the mail by sled dog in winter and packhorses in summer. Photo was taken in 1929 on Jack Creek about 10 miles from Nabesna Mine.

January 28, 1929

My dear Mr. Wood:

        The enclosed petition, accompanied by a letter from Mr. Carl F. Whitham, an engineer who has been working in Alaska in the Copper River Valley, will speak for itself. I have had a conference with Mr. Whitham and am much impressed with the urgency of this need of a small day school and community center for these people as he puts it up to us. I have explained to Mr. Whitham, however, that that part of Central Alaska is, according to a comity agreement subscribed to by our various denominations and approved by the Home Missions Council in 1919, territory in which the Protestant Episcopal Church works and that his appeal should be to you. I told Mr. Whitham further that I should be very glad indeed to introduce him to the Episcopal Board, but since he will be in the city only a few days longer, and I am leaving tonight for a meeting in Cleveland, I am introducing him by letter instead. Anything which you can do for Mr. Whitham when he calls will be much appreciated by your co-workers for the Kingdom, the Presbyterians.

                                                        Very sincerely yours

Enc.
ERV:MC
Dictated by Miss Voss but
signed in her absence

                                                  Edna R. Voss
                                                   Secretary

Mr. John W. Wood
Board of Missions of the Protestant Episcopal Church
New York, New York

Whitham actively worked for roads in eastern Alaska and did so believing Alaska could not progress without roads. During his effort to get roads built he also lobbied for schools in the Copper River area.

a comity agreement approved by the Home Missions Council of 1919, the Copper River area was an area in which the Protestant Episcopal Church works. She introduced Whitham by a letter of January 20, 1929, to Mr. John W. Wood of the Board of Missions of the Protestant Episcopal Church, New York. With the help of others schools were eventually established.

Whitham had an impact on the lives and economic well being of local Indians beyond that provided by roads and schools. Those few who had contact with him and are alive today remember him favorably. Indians from Batzulnetas worked for Whitham wrangling horses, cutting trails, running pack trains, and working with him at what would become Nabesna Mine. Their descendants still live in the area or call it home. One is Angus DeWitt,[49] and another is Wilson Justin.[50] Nearly every family member of the John's, the Joe's, the Bell's, the Charley's and many others in the villages of Chistochina, Slana, Twin Lakes and Mentasta have a grandfather or relative who worked for Whitham or his Nabesna Mine.

When looking back, what is really interesting is that Carl Whitham put so much effort into developing the Nabesna area before he found anything out there that would benefit him. However, that was about to change.

---

[39] President Roosevelt appointed Army Captain Wilds Preston Richardson as the Commission's first president in Jan. 27, 1905. A West Pointer, 1884, he fought the formidable Apache Indians in Arizona, and taught at West Point. He knew Alaska and near the turn of the century spent time with Capt. P.H. Ray at Valdez and Eagle. Capt. Richardson had big ideas commensurate with his bulk. He weighed about 300 pounds. He was a colonel when World War I broke out. He resigned from ARC and in 1917 served in France as a brigadier general.

[40] The same is probably true today.

[41] Dan Sutherland was born in Cape Breton, Canada in 1869 and came to Alaska in 1898 at age 29. He mined in the Nome, Iditarod, and Fairbanks areas. He was a U.S. marshal for a short period and was elected to the first Alaska Legislature in 1912. In 1920 he took his seat as Alaska's Delegate to Congress. Before Statehood in 1959, Alaska had only a non-voting Delegate to Congress. He knew Presidents Harding, Coolidge, Hoover and many influential senators.

[42] The Territorial Road Commission was established after passage of the Alaska Organic Act of 1912, which gave Alaska a territorial legislature. Members of this commission were prominent Alaskans appointed by the governor. The commission was chiefly advisory to the legislature and ARC.

[43] Brooks, Alfred Hulse. *Blazing Alaska's Trails*. Univ. of Alaska, 1953. The present Richardson Highway was named for Gen. Richardson in the 1920s.

[44] Alaska Road Commission members generally traveled to Washington during the winter when Congress was in session to support ARC's request for funds.

[45] The present concrete bridge on Nabesna Road crossing the Slana River was constructed after the "Good Friday "earthquake of 1964.

[46] Dr. Alan M. Bateman, a world renown geologist and an expert on ore deposits served as a consultant for Kennecott and was responsible for directing the early exploration at the mine that lead to recognizing the favorable structure of the ore bodies. In 1942 he published the college textbook Economic Mineral Deposits that became a classic and a standard college text.

[47] For discussion of plans for a railroad that might have included a branch through the Copper River -Nabesna country see Elizabeth A. Tower's books, *Ghosts of Kennecott, the story of Stephen Birch*, Anchorage, Alaska, 1990, and *Big Mike Heney, Irish Prince of the Iron Trails,* story of Copper River Northwestern Railroad, Anchorage, Alaska, 1988, and *Icebound Empire*, 1996. See also Lone E. Janson (1975) *The Copper Spike*, Northwest Publishing Co. in which she includes a map showing the proposed railroad route (1906-1908) through the Slana-Nabesna country to Eagle on the Yukon River.

[48] Stephen Birch had with Lt. Lowe in the summer of 1898 traveled up the Copper River to the Slana River and crossed the Alaska Range on their way to the Yukon River. Kennecott Copper Co. in the late thirties did enter the region and explored a molybdenum deposit in the Alaska Range a few miles north of Nabesna Road and after World War II explored Orange Hill-Bond Creek copper deposit 12 miles south of Nabesna.

[49] Angus is the son of Lawrence and still lives in the historic Slana Roadhouse at Slana on Nabesna Road. As a boy he worked at Nabesna Mine and other gold mines. He's a lifelong Alaskan and a long time friend of the author. Angus is a former Chief of the Mentasta Indians, an ex-soldier, dozer operator, hunter, core driller to mention a few things he did and did well. .

[50] Wilson was born near Nabesna and in recent years served as a member of the Board of Directors of Ahtna Native Corporation and then as President of Ahtna Native Corporation, both established by the Alaska Native Claims Settlement Act of 1971.

CHAPTER 10

THE BEAR VEIN
The Gold Discovery that Made Nabesna Mine

The early prospectors who organized the Royal Development Corp. had tried to make a mine of the vein they found. It is still there on the mountainside. The stamp mill did not work as the miners hoped. It is still there, too. Those early prospectors grew weary of failure, pulled up stakes and moved on into anonymity.

Each time Whitham returned to that lonely place and looked up at those towering cliffs he must have had a sense that the cold hand of failure had drawn ever closer. All too well he knew that of the tens of thousands of prospectors who searched throughout Alaska in that era only a few found anything worthwhile.[51] Closer still, he knew of the failures right here on White Mountain.

Then on an August day in 1926, he searched higher on White Mountain than ever before. High up above the towering cliffs and beyond the sheep cave he walked onto a tundra-covered ridge. On that wind-swept place he discovered one of the richest gold bearing veins ever reported in Alaska.

According to Whitham he picked up a piece of broken quartz and looked at it. Half of the broken quartz was gold! A big grizzly, he wrote, that had dug for a ground squirrel, flung tundra, dirt and broken quartz pieces across the ridge where he stood. He was looking with disbelief at those quartz pieces.

He guessed that he was standing over a gold bearing vein running below his feet on that tundra covered ridge. It was there, all right. Out of it grew Nabesna Gold Mine.

He called it the "Bear Vein". Parts of it assayed 200 ounces gold per ton of ore and more. Today, a common 55-gallon fuel barrel filled with that ore would weigh about a ton and hold about $50,000.00 in gold. It was not so in

Present day photo showing where Whitham in 1926 with the help of a hungry bear discovered the "Bear Vein." The discovery site is the grass-covered saddle to the right of the highest building.

1926 when the price of gold was $20 per ounce. It surely was enough, though, to make Whitham's hair stand on end thinking about what he might have.

As August slipped into September of 1926, fresh snow blanketed the peaks of the Wrangells and the Alaska Range. Summer was ending. However, even with the shortening days he still had time to climb every day that several thousand feet up White Mountain. There with pick and shovel he stripped tundra from parts of the vein.

In mid-October when his drinking water froze solid, as snow swept across and blanketed interior Alaska, he left on a saddle horse with a packhorse in tow. As he rode the trail below those great white cliffs he must have known he had found what he had been searching for all those years and over all those lonely hard miles.

Various reports describe the Bear Vein as a gold-bearing flinty, opaline-like quartz vein three to ten feet wide. This rich vein trended southwest and dipped five to 30 degrees west into the mountain.

Carl F. Whitham, in this 1927 photo, is standing where he discovered and trenched the Bear Vein in the late summer of 1926. The vein was as wide as 10 feet and some sections assayed 200 ounces gold per ton and twice that in silver. The discovery site is 2000 feet above the old "White Mountain Cabins."

In the spring of 1927, he was back on his mountain and hired Steve Kansky to help strip more overburden from the Vein. Later that season Whitham and Kansky packed a homemade dry rocker up to the site. In the American West, miners used dry rockers when they discovered rich gold bearing veins where there was little or no water, just as high up there on the side of White Mountain.

Breaking the rich ore from the vein with pick and sledgehammer they used the rocker to separate the heavier gold from the lighter quartz. The raw gold they recovered paid Whitham's 1927 expenses.

The vein on the surface was rich, but was it just a "flash in the pan?" To find that out Whitham would have to sink a shaft along side the vein. That fall, knowing what he must do next season, he and Kansky left White Mountain. After traveling to the coast, Whitham took the steamship to Seattle.

He and Marie spent the winter together, uplifted by the new prospect. They traveled to New York City and Washington, D.C. There he continued his almost ceaseless efforts to interest Washington in more roads in eastern Alaska. He also spent time writing. He had become a regular contributor of Alaska mining news to the well-known publication then, as today, *Engineering and Mining Journal* (E&MJ). A.H. Hubbell, Associate Editor of E&MJ, in a letter of September 27, 1927 wrote Whitham, "Am pleased that

The Bear Vein was rich enough that when crushed by a sledgehammer raw gold could be separated from quartz by means of a homemade rocker. Whitham was able to pay his 1927 expenses from gold recovered by the rocker. Man holding the rocker is Steve Kansky.

our editorial "Alaska and the Prospector" interested you. As a matter of fact I had you very much in mind at time of writing it."

Whitham's friend, Ike P. Taylor, Ass't Chief Engineer, Alaska Road Commission had heard news of Whitham's discovery and wrote in November 1927,

> "Major Elliott has just arrived in Alaska, as successor to Colonel Steese. He will return to Washington to defend the estimates before Congress. He read your letter and will be very glad to have any assistance that you may be able to give him, specially information concerning the possibilities in your country."

Funding for roads turned out less than hoped for and in a March 27, 1928 letter to Whitham, Dan Sutherland, Delegate to Congress wrote:

> "The amount was finally fixed by the Commerce Committee at $825,000.00, (for all roads in Alaska) so we are $175,000.00 shy of last year. I believe it was the intention of Major Elliott to do considerable work on the road (meaning road to Slana) if he had received the million dollars . . ."

In the spring of 1928, Whitham took dynamite, fuse and caps with him to

ENGINEERING AND MINING JOURNAL

JOSIAH EDWARD SPURR
EDITOR

McGraw-Hill Publishing Company, Inc.
Tenth Avenue at 36th Street
New York, N.Y.

Sept. 27, 1927.

Mr. Carl F. Whitham,
Nabesna Camp,
Chisana, Alaska.

Dear Mr. Whitham:

    I was very glad to hear from you in your letter of Aug. 14 and am pleased that our editorial "Alaska and the Prospector" interested you. As a matter of fact I had you very much in mind at the time of writing it.

    The news you send is welcome and I have turned it over to Mr. Keiser who is handling our news. I am sure that he will be interested in it also.

    Any information that you can send us in the future that we can turn to editorial advantage regarding the need of Alaska for better transportation will be welcome. I am glad to hear that you expect to be in New York this winter and trust that you will find it possible to call upon us, as you say. With kind regards, I am,

                              Yours very truly,

                              A. H. Hubbell
                              Associate Editor

AHH/EEE

Whitham made a continuous effort to make the public aware of and interested in eastern Alaska. He submitted articles to the *Engineering and Mining Journal* and other magazines and journals. The *Engineering and Mining Journal* is still published today.

In 1929 Whitham with the help of the Batzulnetas Indians hauled supplies in by packtrain and the supplies included dynamite to sink the shaft.

Nabesna. Winter was waning when Whitham rode back to Nabesna. Kansky was ready again to work at White Mountain.

They started the shaft where they'd stripped the vein of overburden the previous summer. They had to backpack more equipment up the mountain than they ever needed before. That was like climbing several Empire State Buildings everyday with a hundred pounds on your back, working all day, then climbing back down only to do it again the next day. An old timer estimated a third of their work was taking care of the horses and fixing their grub. Another third getting supplies and packing up the mountain. That left only a third of their working hours sinking the shaft.

They dug that shaft into what is actually a tough marble. The shaft was five feet by five feet and only large enough for one man to hold a two to four foot long drill steel in one hand and with the other drive it down with a 10-pound sledge. It's called "single jacking". They continued this way until the drill holes were about four feet deep. When six holes were completed in the bottom of the shaft, Whitham or Kansky, whoever was doing the drilling, loaded the holes with sticks of dynamite. In certain sticks of dynamite they pushed in a blasting cap, and inserted in it a black powder fuse. The fuse had to be cut long enough so that when lighted–called "spitting"–the man in the hole had time to climb up the shaft cribbing and out the collar. Otherwise bad things happened.

They did the blasting at the ends of their days so that by the following mornings most of the dynamite smoke had drifted out. Then Whitham or Kansky climbed back down and mucked the broken rock into a bucket while the other hoisted it up by a homemade windlass.

As they deepened the shaft, they cribbed the walls with logs to prevent caving. There weren't any trees up there so they cut the logs in the valley and packed them up. Whitham collected samples every few feet from the wall and after crushing the sample at the cabin at night, he panned it. By weighing the gold saved in the pan he could estimate the ounces of gold per ton in the ore. When they had sunk their shaft nearly 50 feet into the solid rock Whitham was sure the vein continued to depth.

Whitham and Kansky sunk a 5'x5' shaft by hand, called "single jacking." They dynamited and mucked out the broken rock by a windlass. Whitham is at the windlass.

He spent the winter of 1928 and 1929 with Marie in Seattle. That winter he continued to lobby his friend Dan Sutherland[52] to push congress for more funds for the Alaska Road Commission so they could improve the 70-mile long trail from the Richardson Highway to Slana River. He didn't get much help for the last section of the trail between the Slana River and Nabesna.

The U.S. Army Signal Corps operated the telegraph line along the 450-mile long Eagle Trail between Valdez and Eagle on the Yukon River. Early on, Whitham tried to get the Signal Corps to string a line to Nabesna. He wrote friends that he didn't think he'd have much luck getting the line but if he did the Alaska Road Commission would have more reason to improve the trail. He was right. He didn't have luck getting the telegraph line. However, several years later the Army would string a telephone line to

Nabesna. He was wrong, though, in thinking the telephone line would raise official interest in the Nabesna Road

In the spring of 1929, Whitham improved the switchback trail from the valley to the Bear Vein. He prepared a portal site on the ridge 100 feet below the outcrop he had exposed and sunk the shaft on. From that portal site he would drive a drift into the ridge for a hundred and fifty feet to explore the Bear Vein underground below the outcrop and prepare it for mining.

In July, Whitham's need for cash overcame his interest in his own prospect. And an opportunity came from his friend D.C. Sargent. Sargent was manager for the Alaska Nabesna Corporation, a company formed by east coast and Washington, D.C. investors who owned Orange Hill copper prospect. As already mentioned, it is some 12 miles south of White Mountain.

Sargent hired Whitham to drive an exploration drift at the Orange Hill prospect. Whitham hired several miners and using his own saddle and packhorses hauled the miners and supplies to the prospect. To get to Orange Hill Whitham had to ford the glacier-fed Nabesna River.

The company paid for the mining work but refused to pay Whitham for a map and report he understood they had requested of him. In a letter to the company's president Whitham reviewed the instructions given him then ended, "I feel that the lesson I have learned in this deal with your company will be very valuable to me in the future." Whitham said this was the only time "he was beat out of money owed him." He didn't blame Sargent and they remained good friends.

Word spread of the rich Bear Vein discovery. Interested parties made offers to buy the property. However, Whitham set about to develop it himself. That would take money he didn't have, and by the fall of 1929, when he left Nabesna for Chitina, he had decided what he had to do.

In Chitina on October 25, 1929 with help of friends he incorporated Nabesna Mining Corporation under the laws of the Territory of Alaska. Whitham organized the first stockholders meeting on October 28th at 7 o'clock in the office of the Alaska Road Commission in Chitina. He had no office and since several of the new stockholders worked for the Alaska Road Commission that office was a convenient place for the company's organizational meeting. Today, private individuals meeting in a public office to organize a private for-profit company would raise no end of howls. Times were simpler then. Whitham assigned all his White Mountain mining claims to the new company and received 50% of the issued stock. The newly elected directors elected Whitham president. He would remain so for the mine's entire operating years.

(Authentication)

A RECORD
of the
ORGANIZATION AND INCORPORATION
of
NABESNA MINING CORPORATION

A CORPORATION

duly incorporated under the laws of the territory of Alaska

on the twenty fifth day of October A. D. 19 29 ,

in the town of Chitina

county of , territory of Alaska

    IN WITNESS WHEREOF, WE, Carl F. Whitham, R.J. Shepard, Thos. S. Scott, D.H. Kelsey, A.E. Moore, John Coats, and Frank Shipp

being all the original subscribers to the articles of incorporation of said incorporation, for the purpose of the adoption and identification of this book, with the records contained therein, as the original record of said corporation, have hereunto subscribed our names and caused the corporate seal to be affixed this 27th day of December A. D. 19 29

Carl F. Whitham
R.J. Shepard
Thos S Scott
D H Kelsey
A. E. Moore
John Coats
Frank Shipp

(To be signed by all the persons organizing the corporation.)

In the fall of 1929 Whitham left Nabesna on horseback for Chitina. With the help of Chitina friends he organized "Nabesna Mining Corporation" on October 25, 1929 and raised $5,000 from local persons needed to drive the drift.

That year proved a turning point for Nabesna as well as for Nabesna Road and in December of 1929, Dan Sutherland wrote hopefully:

> "The road appropriation (for Alaska) for this year, as you probably know is $800,000.00 but as there will be considerable money released from the main Richardson Highway and the Ruby-Poorman project I believe Major Elliott will do a lot of work on the Gakona project." (Gakona project was Nabesna Road).

---

[51] Mineral rich as the tens of thousands of square miles of the Wrangell Mountains and Eastern Alaska Range is said to be, there has only been two commercially producing lode mines in all that area, Nabesna Gold Mine and the Kennecott Copper Mines.

[52] Dan Sutherland served as Alaska's non-voting Delegate to Congress until 1930 and was now in his last term. He had been campaign manager for James Wickersham when the latter was a Delegate to Congress. Wickersham was a well-known Alaskan and came to the Territory in 1900 when appointed as Federal judge at Eagle City. Resigning the judgeship in 1907, he ran for and was elected as Alaska Delegate in 1908. In 1920, he stepped down and convinced Sutherland to run for Delegate on the understanding that Sutherland would support his appointment for governor of Alaska. A fight broke out in Congress over Wickersham's appointment to governor. Sutherland did not support him. President Harding offered the governorship to Sutherland but he refused it. Sutherland stepped down in1930 as delegate and Wickersham ran and was elected, see Evangeline Atwood, (1979) *Frontier Politics*, Binford & Mort. Judge Wickersham seems not to be one to "forget and forgive" for in his book *Old Yukon. Tales-Trails-Trials* published in 1938 he never mentions Sutherland. Whitham and Sutherland were personal friends and while Whitham once supported Wickersham he didn't after Wickersham gave a speech in New York City that supported Ben Eielson's argument that Alaska needed airports more than roads. Eielson was a well-known Fairbanks flier (Eielson Air Force Base was named for him) and represented an aircraft company. Whitham countered in the New York Times saying, "airplanes could be as thick as gulls but Alaska needed roads." Dan Sutherland, a few weeks short of his 86th birthday died March 23, 1955.

CHAPTER 11

The Beginning of Nabesna Gold Mine

Nabesna Mining Corporation was in place. Now Whitham had to raise the money to drive the Bear Vein drift. The work, he estimated, to drive the Bear Vein drift, also called the 100 level, would cost about $5,000.

He called this first drift the "100 level" but more often miners called it the "Bear Vein drift." The portal site he selected was 100 feet down the slope from the outcrop. Thereafter, he labeled his portals according to the number of feet below the outcrop. Thus, the first drift was 100 feet below the Bear Vein outcrop and the next lower drift was 150 feet below the 100 drift, which was 250 feet below the outcrop and thus called the 250 drift.

Whitham never had much cash, but he had horses and tools and was never a down-and-out-prospector. Friends respected him for that. Whitham was a dreamer but also a pragmatist and he never seemed to mix the two; it took a dreamer to keep searching and a realist to know what to do when he found what he had been searching for. He would not have been out there for 17 years always looking had he not been the dreamer. But once he made his discovery, he made hardly a false move in developing his mine.

He raised the $5,000 from the private sale of stock at twenty-five cents per share to friends in Alaska and Seattle. With his new capital he purchased supplies in Cordova and he hired four miners. Then he bought horses in the nearby community of Kenny Lake between Chitina and the Richardson Highway. On his way to the mountain he bought two more horses from his friend DeWitt at Slana Roadhouse.

Snow still hung on the ground in April 1930 when Whitham and his crew improved the switchback trail up the mountain to the Bear Vein site, a total distance of some 3000 feet. Packhorses could go about two-thirds

NABESNA GOLD

Alaska Agricultural College and School of Mines
In Cooperation with
U. S. Bureau of Mines, Department of Commerce

College, Alaska

July 3, 1930.

REPORT OF ASSAY

On samples received from Nabesna Mining Corporation, Chitina, Alaska.

Assay No.	Mark on Sample	OUNCES PER TON		Value Per Ton	PERCENTAGE OF			
		Gold	Silver		Lead			
5759	No. 1	18.33	37.60	$381.64	2.86			
5760	No. 2	11.40	17.20	234.88	2.11			
5761	No. 3	22.38	35.40	461.76	4.67			
5762	No. 4	252.23	119.90	5292.56	47.12			
5763	No. 5	5.04	4.20	102.48	0.12			
5764	No. 6	2.57	5.60	53.64	0.08			
5765	No. 7	3.31	8.20	69.48	0.34			

The values per ton, as given above, do not include the value of the lead.

Assayed by,

Paul Hopkins

Paul Hopkins,
Associate Anal. Chemist,
U. S. Bureau of Mines.

Total charges for above assays............ $14.00
Amount received from sender.............

Copy

In the spring of 1930 Whitham hired 4 miners to drive a drift to explore the Bear Vein. He called it the 100 level drift. It was 100 feet below the Bear Vein outcrop, but miners called it the "Bear Vein Drift." He sampled the vein at 5 foot intervals as the drift was driven. Some samples were sent for assay to Alaska Agricultural College and School of Mines–now the University of Alaska Fairbanks. Others were sent to Seattle. The assays show the high grade nature of the Bear Vein.

of the way up. From there it is too steep, so miners backpacked the rest of the way.

The miners portalled the Bear Vein drift, the "100 level," on the slope below the outcrop. Whitham set up a tent camp at the site, 2000 feet above the valley. That's where he and his men lived that summer. The drift was five feet wide by seven feet high. As the miners drove deeper into the mountain, Whitham sampled and measured the vein every 5 feet. He sent samples for assay to Alaska Agricultural College and School of Mines at Fairbanks, later the University of Alaska, and to Art Glover in Seattle. Whitham also crushed and panned samples as he had when sinking the shaft and made estimates of ore grade.

He had no air compressor and his crew drove the drift by "double jacking." One miner held a piece of drill steel and the other hammered it into the rock with blows from a 20-pound sledge. Misses were few but bloody. After they drilled a "round," meaning a dozen or more holes four to six feet deep in the face of the drift, the men loaded the holes with dynamite and blasted. They hand mucked the broken rock into a homemade wheelbarrow, pushed it to the portal and dumped it in either a waste pile or an ore stockpile. It was crude and primitive, but all across the American West hardrock miners drove tens of thousands of feet of underground workings this same way.

Years later someone asked, "How often up there did the miners working under such dirty conditions and, with water being so scarce take baths." Whitham just shrugged, "Never" and added "They didn't take baths when in town anyway." This, of course, was not true but made a good story. The miners divided up and worked two 10-hour shifts every twenty-four hours. Daylight for most of the summer was at least 18 hours a day so that about the only time it was dark was when the miners were underground in the drift. Underground the miners used carbide lights. They advanced the drift about 150 feet that summer. It was hard work and a rough life-style. That was just the way things were done back then.

Whitham found himself at another crossroads. The 100 level or Bear Vein drift was in high-grade gold ore for almost its entire length. In September 1930 Whitham ended this phase of the work. He paid off the miners and left Nabesna. The next phase would be something unlike anything he had attempted before. He left Nabesna that fall knowing he had to raise at least a $100,000 to develop the mine and build a mill.[53]

With help of Seattle friends he listed Nabesna Mining Corporation on the newly formed Curb and Mining Exchange. The Curb and Mining Exchange was affiliated with the Seattle Stock Exchange located in the

Merchant Exchange quarters of the Chamber of Commerce building. The sale of stock quickly raised the needed money. This was no small feat in this era. The New York Stock Exchange had collapsed and the United States and the World was entering the years of the Great Depression.

---

[53] This would be equivalent to about $1 million at year 2002. This was during the start of the Great Depression when the stock market collapsed, but because his friends had faith in him and his mine, the money was timely raised.

CHAPTER 12

LITIGATION

Every hard working miner has a litigation story–or several. For every person who goes out and digs in the ground and works like a dog, there is at least one jackal lurking about trying to figure out how to get away from the miner by some dodge whatever the miner earns by honest toil. Doubtlessly, many of the California Gold Rush successes were jackals. Because of this unfortunate inevitability, probably no successful gold mine has ever not weathered a lawsuit. Nabesna is no exception.

In this case it was a colorful character named James J. Godfrey of New York City. One may recall Godfrey came into Whitham's life years earlier when Whitham worked for him at McCarthy. In sum, Godfrey claimed that by a grubstake agreement with Whitham he owned 50 percent of the White Mountain mining claims on which Whitham had his gold mine. In December 1930, James J. Godfrey filed a lawsuit against Whitham.

Godfrey was a graduate-mining engineer of Columbia School of Mines in New York City. Worse, he was a lawyer too. He had worked in American and Canadian mines and many prominent mining men of the day knew him well. Godfrey was particularly well known in the Territory of Alaska. In 1907 he organized and was president of The Mother Lode Copper Mines Company of Alaska located on the opposite side of the ridge from Bonanza and Jumbo Mines. The Alaska Syndicate, a partnership between the Guggenheims and Morgan Bank, controlled the latter two mines. Kennecott Copper Corporation became organized and eventually acquired them.

According to affidavits filed by Godfrey in his suit against Whitham, Godfrey stated that he sold his company (Mother Lode Copper Mines Company of Alaska) in 1919 to the Mother Lode Coalition Mines Company controlled by Kennecott Copper Corporation. Godfrey's company, The

Mother Lode Copper Mines Company of Alaska went out of business. During this period Godfrey acquired a financial interest in Kay Copper Corporation in Arizona and became its president.

As already mentioned, Carl F. Whitham worked at The Mother Lode Copper Mine in 1913 and met Godfrey. Months after working there Whitham quit his job in Godfrey's mine, joined the Shushana gold rush and worked his placer mine until he enlisted in the U.S. Army during World War I. At war's end he was honorably discharged in the winter of 1919 on the east coast, and Whitham, being an affable person who seemed to keep track of all acquaintances, visited James Godfrey in New York City and renewed old acquaintanceship.

Leaving the City, Whitham, as we know, journeyed to Hot Springs, Arkansas, where he married Marie. For the next five years, he and Marie lived in Alaska at Chisana and White Mountain until 1924, when he and Marie purchased a home in the Seattle area. Whitham returned to Alaska and continued working and prospecting. Then in the fall of 1925 he and his wife visited New York City,

There, as was his way, he visited Godfrey. He told Godfrey, according to Whitham's affidavit, about the White Mountain claims he had staked in 1922. Godfrey was not interested in gold, but was interested in the silver-lead prospects Whitham knew about in the Slana-Ahtell region some 50 miles west of White Mountain and some 10 miles from the Slana Roadhouse. Before he left New York he and Godfrey on January 30, 1926 entered into a grubstake agreement that included staking the silver-lead prospects in the Ahtell-Slana area. Whitham would stake and explore these and others prospects, and Godfrey would pay the expenses. The fruits from the claims staked would be shared 50-50. The Seattle newspaper *Alaska Weekly* on May 13, 1927 said that Whitham had returned from the East on his way back to Alaska. He and Godfrey, the paper told, had joined forces to explore silver-lead prospects in the Ahtell-Slana area. During the summer of 1926 Whitham staked about 30 silver-lead claims in the Slana-Ahtell area. Whitham, while working at Nabesna, hired men to work on the silver claims in the Slana-Ahtell Creek area. He billed Godfrey, but Godfrey did not pay.

Back in New York things were not going well for Godfrey and on May 7, 1927, the *New York Times* headlined: "Bank Lent $20,000 Got Defunct Stock". The column went on to tell that James J. Godfrey had borrowed $20,000 from American Exchange-Irving Company, a Wall Street Investment Bank, and gave as security 5,500 shares of The Mother Lode Copper Mines Company of Alaska. The problem was, as the bank was to learn, Godfrey had sold the assets of The Mother Lode Copper Mines

> **NEWLY FOUND LEAD CAMP AT WRANGELL MT.**
>
> Carl F. Whitham, Who Has Properties There, Taking Small Crew to Start Preliminary Work
>
> **JUST BACK FROM THE EAST**
>
> James J. Godfrey, of Mother Lode Fame, Is Concerned With Whitham in the Properties in New Camp, Which Is at the Headwaters of Tanana and Copper Rivers
>
> Carl F. Whitham, pioneer mining man of the Copper River valley region and the Shushanna placer diggings, arrived in Seattle this week, on his way to Alaska, after an extended trip through the East and in the southern states. He is leaving on Saturday's northbound steamer, and will take a small crew of men with him to his mining claims in the interior.
>
> "Mr. James J. Godfrey is associated with me in mining properties up there," said Mr. Whitham, "and it is our intention to make all preliminary preparations possible this summer for the carrying on of an extensive development program next season. We have silver-lead properties located in the newly discovered North Wrangell mountain lead district, at the headwaters of the Tanana and Copper rivers. Mr. Godfrey, who is well-known in Alaska, where he spent many years, and developed the Mother Lode property, now controlled by the Kennecott Copper Corporation, into a productive mine, is now living in New York City, and is president of the Kay Copper mine in Arizona, which will be brought into production soon, as plans have been made, and all finances arranged for the installation this summer of a mill around one thousand tons per day capacity to treat the large tonnage of ore at present blocked out.

> **BANK LENT $20,000, GOT DEFUNCT STOCK**
>
> J. J. Godfrey Is Sued to Force Exchange for Shares in New Corporation.
>
> **INJUNCTION ALSO SOUGHT**
>
> American Exchange-Irving Company Seeks to Prevent Disposing of Mining Issue.
>
> An unusual financial transaction, in which a bank alleges that it became the holder of stock in a defunct corporation as security for a loan of $20,000, was disclosed yesterday in the Supreme Court when the American Exchange-Irving Trust Company applied for an injunction restraining James J. Godfrey from disposing of any stock of the Mother Lode Coalition Mines Company pending the suit, and compelling him to exchange 8,500 shares of stock in the Mother Lode Copper Mines Company of Alaska for 13,600 shares of Coalition stock.
>
> Mr. Godfrey has been President of the Mother Lode Coalition Mines Company and the Kay Copper Corporation. The shares of the Kay Corporation were stricken from the curb list on Feb. 17 after the price of the shares had dropped from $1.25 to 25 cents a share. The papers filed yesterday said that representatives of Mr. Godfrey explained that he had been too busy watching the affairs of the Kay Copper Corporation to devote any time to the matters involved in the suit.
>
> **Loan Obtained a Year Ago.**
>
> The petition by Howard Marshall, Vice President of the bank, alleged

James J. Godfrey, a mining engineer and lawyer, organized the "Mother Lode Copper Mines Co. of Alaska" in 1907. Whitham worked there in 1912 and met Godfrey. Following the "stock market crash" in 1929, a Wall Street bank demanded Godfrey pay $20,000 the bank had loaned him. Learning Whitham had raised money to develop Nabesna Mine, Godfrey tried to borrow money from him. Whitham declined and Godfrey sued, saying he owned 50% of Nabesna claims. The court found Godfrey had no such claim to Nabesna.

Company of Alaska in 1919 to Kennecott and the company had gone out of business.[54] The stock given to the bank as security, so the investment bank said, was worthless. As to why he put up the worthless stock, the *Times* reported Godfrey said he'd been too busy with his Kay Copper Company mine in Arizona to pay attention to the deal with the investment bank.

Things weren't going too well in Arizona either. The price of copper had dropped and shares of Kay Copper Company had plunged from $1.25 to 25 cents per share. The New York Stock Exchange struck the shares from trading on its exchange. Back in 1919, when Godfrey sold the Mother Lode Copper Mines Company of Alaska to Kennicott, Godfrey had received shares in the Mother Lode Coalition Mines owned by Kennecott. The bank demanded he put up those shares. Godfrey stalled. The bank filed a complaint in the Supreme Court of New York, seeking an injunction restraining Godfrey from disposing of any of his Coalition stock. Godfrey held 13,000 shares of the stock. The bank said the stock was worth $4.25 per share. Plainly, Godfrey was not broke, but because of the bank's lawsuit he could not sell or use the stock for collateral. He was solidly on the hook for $20,000 to American Exchange-Irving Co. and needed money

Godfrey learned in early 1930 that Whitham had organized Nabesna Mining Corporation and had raised money. He made inquiries into Whitham's activities. The money he believed Whitham held as a result of those activities greatly interested Godfrey. He wired Whitham asking to borrow the money he needed. Whitham wired back saying that the money from the sale of Nabesna Mining Company stock went to buy equipment. Godfrey apparently didn't believe Whitham. After repeated rejections from Whitham, Godfrey filed a complaint in December 1930 in the Superior Court of the State of Washington styled: *James J. Godfrey, Plaintiff,* vs. *Carl F. Whitham and Nabesna Gold Mines Corporation, a corporation.* Court Number 237604.

The lawsuit claimed Godfrey owned 50 per cent of the White Mountain claims and asked the court to enjoin Whitham from disposing of any of his Nabesna Mine stock until ownership of the claims was settled. Godfrey alleged he had entered with Whitham into a verbal or oral grubstake agreement in New York allegedly in 1922 concerning the White Mountain claims! The written grubstake agreement of January 30, 1925 allegedly formalized the alleged oral grubstake agreement about White Mountain. The written agreement, therefore, according to Godfrey, required that all claims staked by Whitham to be shared 50-50 with Godfrey. In effect Godfrey claimed he owned a 50% interest in White Mountain because the written agreement was a record of his 1922 oral agreement with Whitham and would include White Mountain.

Whitham rejected Godfrey's claims. Each side filed a number of affidavits. Whitham obtained a goodly number from Alaska and Seattle friends including John Muller, President of First Bank of Cordova; Austin "Cap" Lathrop, a prominent Alaska businessman and reported to be Alaska's first millionaire;[55] Earl Knight, publisher of the *Alaska Weekly,* and others attesting to Whitham's honesty. Godfrey responded in kind with well-known mining persons and friends in New York area.

This was at the very time when Whitham was busy with development at Nabesna Mine. Whitham's absorption in Nabesna is demonstrated when his Seattle lawyer, Edward F. Medley, pointed out he was handicapped in trying to get word to and from Whitham at the mine. He made this point when he was providing reasons for causing the court to allow extensions. In one letter he advises Whitham that a letter from Seattle by boat to Cordova and then delivered to him at Nabesna and then the reply back to Seattle took about six weeks; whereas, correspondence from Godfrey to his Seattle lawyer took only about 10 days by train.

In the end Whitham successfully defended. He proved he wasn't close to New York City in 1922, when Godfrey said they made a verbal grubstake agreement and could not, therefore, have had that alleged verbal agreement with Godfrey.

Whitham entered the written grubstake agreement signed Jan. 30, 1926, four years after he had staked his mining claims at White Mountain. Apparently, Godfrey found out when Whitham staked his White Mountain claims, probably from the voluble Whitman himself, and tailored his story about the oral 1922 grubstake agreement so that it predated the 1922 staking. However, because in fact Whitham staked the White Mountain claims before he talked with Godfrey about prospecting elsewhere, the White Mountain claims could not reasonably have been part of the 1926 grubstake agreement, which Godfrey claimed formalized a 1922 deal. Just as importantly, the January 30, 1926, agreement spelled out the area of interest where claims were to be shared 50-50. It specified areas from the Nabesna River east to Chisana and in the Slana-Ahtell region. White Mountain is west of the Nabesna River. The claims at White Mountain being west of the Nabesna River could not have been part of any agreement about efforts east of the river. The claims in the Slana-Ahtell area are 50 miles to the west of White Mountain. The case ended in Whitham's favor.

With this out of the way Carl Whitham could focus on what was really important–the building of a gold mine. And that is just what he did.

---

[54] According to Wm. C. Douglass, destructive snowslides in spring of 1919 wrecked much of the surface installations including tramway and power lines at the Mother Lode Mine. Money was needed to repair the damage and pay for exploration. Kennecott advanced the money and received operating control of the Mother Lode. The Mother Lode was connected underground with the Bonanza Mine and it became a profitable mine. Douglass, Wm. C. 1964, *History of the Kennecott Mines*, Seattle, Washington. According to Douglass, Jack Smith and Clarence Warner discovered the Bonanza and Jumbo deposits (mines) in 1900. After selling their interest they continued prospecting and discovered the Mother Lode. This is one of the few times in mining history two men discovered mineral deposits within several years of one another that would become three important producing mines.

[55] Tower, Elizabeth A., 1991, *Cap Lathrop's Keys for Alaska's Riches MINING MEDIA MOVIES*, Anchorage, Alaska.

WHITHAM TO RUSH PRELIMINARY WORK ON NABESNA DEVELOPMENT

Carl F. Whitham, head of the Nabesna Mining Company, will leave for the interior tomorrow and begin work on freighting twenty-five tons of mining supplies out to the company property on the Nabesna.

Included in this freight is an eighteen hundred foot aerial tram with bucket equipment, that will be installed from the pay ground to the mill location.

The tram is out of the ordinary in that it will span the entire distance without intervening towers. It will operate without power other than is supplied by gravity, with the loaded buckets carrying the empties back up the hill.

About February first Mr. Whitham will assemble a crew of seven or more men and begin work of putting in the tramway, digging the tunnel on the ore vein, putting in ore shoots and bunkers, constructing buildings, and preparing the mill foundation so that all will be in readiness to install the mill the first thing next year.

This mill, which will be in operation early in the spring of 1931, will handle not less than thirty tons of ore a day, and will be modern and efficient in every respect.

The greater part of the freight which Mr. Whitham will take into the mine has been taken out the road from Chitina for 120 miles but the last seventy miles to the company property will be made with teams and bobsleds over the winter sled road.

NEW SALT LAKE WORLD WONDER

NEW YORK, Jan. 22.—A dispatch to the New York Times from Adelaide, Australia, says Cecil Madian, Australian explorer, had verified existence of a vast salt lake which may be one of the wonders of the world.

The lake is known as Lake Eyre, and is in the middle of the sunbaked desert of central Australia. Madian believes the lake contains almost limitless salt, with at least 3,000,000,000 tons in the northern part alone.

Should chemical analysis show potash in the salt, the Times says, the lake would be found to be enormously valuable. The explorer said there probably was no other lake in the world with a surface like that of Eyre, which is covered with encrusted patches of crystal salt, like ice floes on an Arctic sea.

For years cattlemen returning from the central Australian desert brought back stories that Lake Eyre was a bottomless morass and that to set foot on it spelled death. Thousands of cattle have been reported swallowed up in it, and the black bushmen living nearby have feared to touch it.

The explorer says these stories were mere superstitious. He drove miles on the lake's surface with a heavily laden motor truck, camping and collecting salt samples.

In the fall of 1930 Nabesna Mining Corp. shares were sold on the Seattle exchange and $100,000 was raised. The January 22, 1931 *Cordova Daily Times* reported that Whitham was freighting mining equipment to Nabesna

CHAPTER 13

Building Nabesna Gold Mine

A January, 1931, issue of the *Cordova Daily Times* said Whitham had purchased about 150 tons of milling and mining equipment in Seattle and shipped it via Alaska Steamship Company north to Cordova. The article told how he and a crew of seven or eight men would freight the equipment by the Copper River and Northwestern Railway to Chitina and from there haul it up to the Richardson Highway and then to Gakona Roadhouse, the trail head to Nabesna.[56] There teams of horses and a tractor would pull bobsleds the remaining 104 miles to the mine. Whitham did just as the article foretold and arrived in Nabesna in spring to confront a new set of challenges.

His first challenge was housing and feeding a larger crew than he had needed before. The old timers had put up the White Mountain log cabins in 1898 and those cabins became the defining mark for years at Nabesna. And, of course, Marie and he had lived there in the summer of 1921 and 1922. But he had to have more than those cabins. So he built a temporary tent camp roughly where the old town now sits on the lower slope of White Mountain.

By mid-May tremendous effort allowed most supplies and equipment to be on site. Men set up and were operating a sawmill, and trees for lumber were cut nearby the mine. For large timbers they cut trees at Twin Lakes and Jack Lake and horses hauled them to the mine. Twin Lakes is a grayling-inhabited lake and "Sportsman Paradise Lodge" is now located there alongside the lake and Nabesna Road 15 miles before getting to the mine. The heavy beams sawed from those larger trees were used in mill construction and for the aerial tramway towers.

To get equipment and heavy beams to the proposed upper tramway site, horses pulled loaded stoneboats three-quarters of the way up the switch-

The equipment and supplies were freighted on a bobsled pulled by teams of horses and a dozer 104 miles from the trailhead on the Richardson Highway to the mine. The small building in the upper photo is a shelter cabin halfway between the Slana Roadhouse and the mine built by Whitham for stranded travelers.

back trail. Where the trail became too steep for horses, the crew muscled the loads the rest of the way.

Construction of the aerial tramway was the most daunting job Whitham and his crew faced early on. The tram cables had to stretch almost 2,000 feet up the cliffs to a site near the portal of the Bear Vein drift. Whitham, the year before in 1930, had staked out the site where he wanted the upper tram terminal built. Whitham's engineer in Seattle designed the tramway based on Whitham's measurements. The engineer ordered the cables. The Cable Company pre-cut them as specified. They were among the equipment hauled from Seattle and represented some of the heaviest loads. It had to be right.

Tent camp at Nabesna in 1931. White Mountain Cabins were present but not enough to house the larger crew.

Once the crew starting making measurements for the terminal site on the cliffs they found a problem. Whitham, no doubt with alarm, notified the Seattle engineer. The engineer replied by letter of June 24, 1931, "What worries me is how you are going to make 1900 feet of cable reach 1945 ft. plus something for anchoring (the tower) and for the weight box ends."[57]

That indeed was worrisome. But to a crew that wrestled five to six tons up those cliffs by strong arms and backs, anything seemed simple. They had a simple solution. Move the terminal tower to the very edge of the cliff and

Some of the mine crew during early days at Nabesna. They are standing on the lower part of the switch back trail leading up to the Bear Vein.

The upper aerial tram tower at Bear Vein drift was set up. This was the tram that needed 1945 feet of cable but only 1900 ft. was available. Resourceful miners hung the tower over the cliff. Photo about 1931.

hang it over. They were indeed tough men with a tough job and they made it work.

They built each terminal tower 20 feet high and put a 150-ton-capacity ore bin in their middles. When these men finished erecting the upper and lower tramway towers and ore bins, they pulled two 1900-foot long 7/8-inch diameter cables weighing two and one half tons each by hand, side by side up the cliff. Their next task was to rig these huge cables on the tramway towers. At this point the two cables were lying side by side on the ground up the cliffs. The men had to suspend them on the towers and get the slack out so that the cables stretched taut up the cliff and well above the ground.

Whitham and his crew did it this way: The two heavy cables, called track cables, were strung over a sheave wheel that the men had mounted on each side of the 20-foot high upper tower. Next, they took the end of each cable and anchored it in solid rock behind the upper tower.

At the lower tower the men pulled the loose end of each cable over sheave wheels also mounted on each side of the tower. However, unlike the upper sheave wheels, the lower ones could grip the cables and be locked in place from turning. With the upper end solidly anchored in rock, the crew could pull the cables as tight as they wanted. Whitham and his crew tightened the cables this way: They built two log boxes about the size of a present day

Portal of Bear Vein drift. The miners dumped waste rock for a rail bed from the portal to the tram terminal on the cliff. Photo about 1932.

Volkswagen bug and clamped one box on the loose end of each cable. They called the log boxes "weight boxes." The boxes were now hanging on the end of each slack cable stretching up the cliff to the upper tower. The crew now filled each log weight box with rocks. As they added more rocks slack was taken out of the cables. When about seven tons of rocks were in each weight box the cables stretched taunt up the cliff to the upper tram terminal. To keep the cables taut over the years of use and stretching, mine employees added rocks from time to time to the boxes.

The crew's next job of creating a functioning tram was to hang below each heavy track cable an ore bucket with arm attached to a roller. They hung one bucket below the cable at the lower and the other at the upper terminal. The men then looped a smaller, one-half inch diameter 4,000-foot long continuous cable that stretched between the two towers. The cable looped over horizontally mounted sheave wheels at the lower and upper terminal. The crew then clamped that cable to the bucket hung on the heavy track cable at the lower and upper terminal. The smaller half-inch diameter cable is called the "tracking" cable. When the miners loaded the upper ore bucket with 600 pounds of ore and released the brake, the bucket rolled down the heavy track cable and as it did the " tracking" cable pulled the empty bucket up from the lower terminal. The aerial tram operated entirely by gravity as its source of power.

The lower tramway terminal was built adjacent to and above the mill with a 150-ton ore bunker. The ore bunker is not yet closed in. The tram bucket is loaded with timber for Bear Vein drift. Miners rode up in the bucket. Photo July 1931.

When the tramway was constructed a weight box was hung on the end of each cable at the lower terminal to insure the cable remained taut for its 2000-foot length up the mountain. About seven tons of rocks were added to the boxes to keep the cable taut. Devils Mountain is in the background.

Whitham's crew trammed a compressor up to the Bear Vein portal along with other equipment to build a blacksmith shop and a shelter room. Those structures remain today. An on-site air compressor enabled the miners to replace double-jacking with machine drilling. Machine drilling could outproduce double-jacking many times over. Those compressed air driven machine drills are called drifters. Others of another type are called stoppers. Whitham insisted all machine drills be fitted with water jackets so water could be forced into the hole through the hollow drill rod and bit. Using water during drilling eliminated rock dust and prevented silicosis. Drilling is faster and less messy without water. But Whitham insisted for the miners' health that they use water and any miner not doing as he was instructed would be fired.

The mine developed rapidly. By the time the tramway was operating another crew had constructed the mill, and it was ready to operate. By mid summer the crew had nearly tripled in size relative to the original crew that Whitham brought with him on his April trip from the coast. He always needed more equipment, men and supplies from the outside.

Airplanes for hauling men and supplies were by 1930 becoming popular in Alaska. Whitham believed they could serve Nabesna. He, therefore, had several men brush out a landing field alongside Nabesna River on a large gravel bar a few miles from the mine. He called it "Nabesna Landing Field." Many well-known bush pilots of the time used it, flying gold out and men and supplies in. Ten years later, only months before World War II

The first section of the mill building was constructed in 1931. The mine sawmill cut all its lumber. The company continued to make additions to the mill over the years of operation.

Whitham had men clear a landing field he called "Nabesna Landing Field" along side Nabesna River. Bob Reeve, Harold Gillam and other well-known bush pilots flew gold out of the landing field. Shortly before World War II the government greatly enlarged the "landing field" to serve warplanes. The military named it "Reeve Field." Caribou are seen in the photo taken in 1931.

broke out, the Federal Government enlarged the field. It became known then as "Reeve Field".

While Whitham and his men mined ore from stopes, cave-like mine workings above the Bear Vein drift that reached up to the original outcrop, Whitham had other miners drop down the cliff another 150 feet. There they started a second portal called the 250 drift.[58] To serve this drift, Whitham and his men constructed a second tram terminal with its own ore bunker at the portal. At the 250 portal miners built a compressor house, blacksmith shop and shelter room. The structures remain and are plainly visible on the cliffs today. But rockslides, snow and wind have not been kind.

The 250 drift had two purposes: one, it was placed on the face of the cliff 150 feet vertically below the Bear Vein drift so that ore between the 250 drift and the Bear Vein drift could be mined and dropped down ore chutes to the 250 and trammed to the mill; two, more of the rich ore would be developed.

In 1932 the mine had still not winterized its living facilities. For that reason Whitham closed the mine in October. Even then there was still only a trail to Nabesna. Part of the crew loaded gear and themselves onto a bobsled that a dozer pulled to the landing field alongside Nabesna River. From there Harold Gillam or Bob Reeve flew the men out to Copper Center.

Before Nabesna Road reached the mine and when the mine closed in October for the winter, part of the crew were taken by sled to Nabesna Landing Field. The miners in the sled are dressed in their good clothes. Other men stayed and took the dozer and sled out Nabesna Trail to the Richardson Highway. There the dozer remained for the winter and in the spring it pulled supplies and materials back to the mine. One or two men stayed at the mine as watchmen.

The same men on the sled in the above photo are standing beside Harold Gillam's plane at Nabesna Landing Field ready to fly out to Copper Center. Gillam, with the white scarf, was based in Fairbanks, and Bob Reeve flew out of Valdez. Devils Mountain is in background.

Recent photo of the "Glory Hole." When the last ore above the Bear Vein drift was blasted down a crater broke up to the surface. Earl Pilgrim a long-time mining engineer and the first professor of mining (1922) at Alaska Agricultural College and School of Mines, now the University of Alaska Fairbanks wrote that the term "Glory Hole" was coined by Western miners when a blasting mistake sent rocks crashing down on miners sending them to "Glory."

Afterwards, the other crewmen loaded onto the bobsled and the dozer pulled it and the men the 46 miles to the Slana Roadhouse. The next day they completed the remaining 60 some miles to the Gakona Roadhouse near the Richardson Highway. The dozer and bobsled were left at the roadhouse to be used again when the mine reopened in April 1933. Old timers concurred it was not a "nice sled ride."

By midsummer of 1933, the company's crew had blasted ore chutes and manways between the 250 level and Bear Vein drift. Miners could drop ore down to the 250 drift where it was loaded into mine cars and hand pushed to the portal and dumped into the ore bin above the tram. The tram operator filled the tram buckets from the chute. Once filled and the brake released, the bucket plummeted down the cable to the lower ore bin above the mill.

The mill crew operated on three shifts 24 hours per day and processed 40 to 70 tons per day. It is interesting to compare this daily production to the total 60 tons processed by the stamp mill in its years of operation.

The miners eventually blasted down the last of the rich ore above the Bear Vein drift to the surface outcrop. The resulting cave was a funnel shaped hole that broke all the way to the surface. The miners called it then, as it's called today, the "Glory Hole." That term, Earl Pilgrim believed, came about from miners in western U.S. mines being killed when a blasting mistake was made and the cave broke through to the surface and rocks crashed down and sent miners to "glory."[59]

Ore from the Bear Vein drift and stopes above it was rich. Some graded as high as 200 ounces gold per ton of ore, but mostly it graded four to five ounces gold per ton.[60] Whitham expected to find that same rich grade as his miners drove the 250 drift deeper into the mountain. Well, it didn't happen.

Instead, his new 250 drift intersected massive, coarse grain pyrite ore grading about one-ounce gold per ton. This lower grade ore was first called "tactite" and later "skarn" ore.

The ore had obviously changed. Some miners described the rich ore of the Bear Vein like that of a roof that slanted into the mountain and lay above the lower grade ore they found in the new drift. The miners believed they came in under that roof.

Whitham was disappointed with the lower grade ore but one-ounce gold/ton is still very good. He didn't subscribe to the "slanted roof" idea and said the theory was full of the same stuff as that in the little house out back with a slanted roof. There's an old maxim in mining that says, "when you're in ore, follow it and don't go wandering off looking for something else." Whitham was of that mind, and he stayed with the lower grade skarn ore.

When asked, " Why haven't you found more of the rich ore?" Whitham's reply was simple: "It just changed."

One man, Phil R. Holdsworth didn't think it was a simple change. Holdsworth came to work at the mine in 1931. He was then a mining engineering student at University of Washington in Seattle. In his years at Nabesna he worked as assayer, mine engineer and later as mill superintendent. He worked at Nabesna until leaving in the fall of 1936, when for the last time he returned to school and got his degree.[61]

He believed the rich gold-quartz ore of the Bear Vein was entirely different from the lower grade skarn ore. The rich ore he believed had deposited millions of years after the lower grade skarn ore. The rich ore was in and controlled by a strong fault system up to 20 feet wide that cut southwest through the mine area. He believed it was a system of several parallel faults that pinched and swelled along their course, called the strike. Holdsworth called it the Bear Vein Fault system. The 250 drift, he believed, had gone under the footwall of the westward dipping Bear Vein Fault (like the miners thought) and by blind luck intersected the lower grade and much older 'skarn' ore, which the mine was now in.

Philip R. Holdsworth, a mining student at the University of Washington, Seattle, started work at Nabesna Gold Mine in 1931. He was the mine's first assayer, later mill superintendent and mine engineer. Likely as not the mine would not have been as successful as it was had it not been for Holdsworth. In the spring of 1937 he graduated as a mining engineer and, after marrying his sweetheart Peggy, left with her for the Philippine Islands. There he became mill superintendent at the Mindanao Mother Lode Gold Mine, one of the large gold mines in the Far East. When the Japanese invaded the Islands in 1941 Holdsworth was commissioned in the U.S. Army. He ordered the mine's pumps turned off and the mine flooded to thwart the invaders. He joined the Army forces fighting in the mountains. Later the Japanese army captured him. He and Peggy spent the war as prisoners. After the war they came to Alaska. Holdsworth worked for the Army Corps of Engineers and then became Commissioner of Department of Mines for the Territory of Alaska. After Statehood, in 1960, he was appointed the first Commissioner of the Department of Natural Resources, State of Alaska. Among his many accomplishments as commissioner he guided the State to select the lands on the North Slope of Alaska that would become the great Prudhoe Bay Oil Field. He retired from public service and died on June 3, 2001.

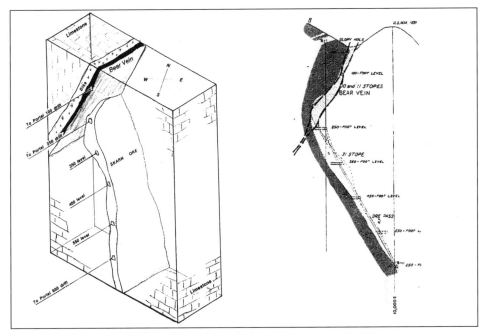

LEFT: The rich Bear Vein dipped north into the mountain. When miners drove the 250 drift below the Bear Vein they struck lower grade skarn ore (1 oz./ton) instead of rich Bear Vein ore. The Bear Vein they believed laid in the mountain like a "slanted roof" and the 250 drift had gone under it. Whitham thought the "slanted roof" idea was full of the same stuff as the little house with the slanted roof out back. Years later he admitted he likely was wrong.
RIGHT: Plate 12, USGS Bulletin 933-B, 1943, is similarly oriented as is the sketch to the left. The heavy dash line superimposed on the USGS cross section shows the Bear Vein and is what Holdsworth believed was the "Bear Vein Fault System." The fault and Bear Vein he believed continued down the dip.

Holdsworth recommended exploration drifts be driven northwest from the 250 drift to find the down dipping Bear Vein Fault. To Whitham that was more of the "slanted roof" idea and a waste of time and money. Many decades later when Alaska was a state and Holdsworth was the first Commissioner of the State of Alaska Department of Natural Resources he remarked half jokingly "Carl didn't much care for a student engineer telling him how to run his mine". Probably true, but in this case, perhaps Whitham should have listened.

Earl Pilgrim,[62] then a mining engineer for the Territory of Alaska examined the mine in 1930 and pointed out the rich ore of the Bear Vein was controlled by a strong west-dipping fault. Asa C. Baldwin, U.S. Mineral Surveyor[63] had also examined the Bear Vein outcrop, and as early as 1929 reported the rich ore was controlled by a strong fault.[64]

Nevertheless, within the lower grade ore the miners found high-grade bodies and some so rich as to be selectively mined and lowered in clean tram buckets to the mill so as not to be diluted by the lower grade ore. However, by and large, from the 250 level down to the lower levels of the mine the ore was massive pyrite-calcite grading about one-ounce gold per ton. As mining continued, rumor persisted the rich ore of the Bear Vein laid unmined in the mountain.

In October 1932, Kennecott Copper Corp., because of low copper prices, temporarily closed its copper mines for the winter. This might have been far-off Alaska, but it was still part of the world economy buffeted by the Great Depression. Its managers closed the copper mines again in the winter of 1933. As a result, the Copper River and Northwestern Railroad announced it would thereafter shut down every year between October and April.[65] Nabesna Mine had used the railroad for freighting from Cordova to Chitina. From there the freight moved over the Edgerton Highway to the Richardson Highway to reach Gakona Roadhouse (trailhead to Nabesna). The closure required the mine to ship all freight during those months from Seattle to Valdez for shipment over the Richardson Highway. Operation of the railroad was expensive even without accidents.

When Thompson Pass was cleared of snow and opened in the spring, the mine used the developing road system instead of the railroad. It proved $10.00 per ton less expensive to truck supplies from Valdez dock to the mine.

Back at Nabesna, Whitham had miners drop another four hundred feet further down the cliffs from the 250 and start the 650 drift. He ordered a third aerial tram constructed at the new portal. A conspicuous rusty stained outcrop, clearly visible from the mill and town and called 'Tower Knob' was the site of the 650 portal. When one climbs to the upper reaches of the mine from the mill and reaches Tower Knob, he knows his misery is just beginning.

A 20-foot wide fault called the Tower Knob Fault cuts the outcrop. Sections of this fault grade one to three ounces gold per ton. Whitham believed that fault continued for at least 1,000 feet southwest across the mine property to the old stamp mill workings.

Besides developing new ore bodies, the 650 drift would allow ore mined from above it and below the 250 level to be dropped down ore chutes where it could be hand trammed by rail cars to the 650 portal and lowered to the mill. Miners built a third tramway to connect the 650 with the mill. Eventually, Whitham put in the 350, 450 and 550 levels between the 650 and 250 levels. Called intermediate levels, there were no portals to the surface; miners accessed them by climbing up manways from the 650 level or climbing down from the 250. As old time miners said, it was pretty dark in those intermediate levels if your carbide light went out.

Unlike many mines in the western U.S. and Canada where ore was developed by vertical shafts and hoisted by headframes to the surface, Nabesna required no shafts. It was able to use gravity to drop ore down ore chutes to lower levels where it could be trammed down to the mill. Nabesna mining was efficient, cheap and smart.

The 650 drift intersected the Tower Knob Fault about 100 feet in from the portal and ore graded one to three ounces gold per ton. Whitham had the miners turn ninety degrees off the 650 drift and drive an adit along the fault to mine the ore. But then he stopped the mining once it became apparent there was not enough rock, (miners call it "back"), between the roof of the adit and the deep gorge on the surface to support the overlying rock. If the roof caved to the surface and into the gorge, the 650 could be flooded. There were other ways to mine the ore and Whitham had the miners continue the 650 into the mountain. Thereafter, the exclusive use of the 650 was in allowing ore mined above it to be dropped down to that level and trammed to the mill.

Before Phil Holdsworth left the mine in 1936 he recommended an outcrop of gold-quartz on the steep mountainside about 600 feet southwest of the glory hole be explored. He believed what he saw there could be part of the southwest extension of the Bear Vein fault system.

In the summer of 1937 Whitham acted on Holdsworth's advice. He had air and waterlines from the 250 portal lain across a wide, steep talus slide to the outcrop. His miners tuneled a drift and Whitham called it the "Nugget Tunnel." The miners intersected the gold bearing fault and followed it for 200 feet underground. The fault was 10 feet wide in places and graded three to five ounces gold per ton. Some five-foot wide sections assayed 35 ozs. gold/ton

The "Nugget Tunnel" surely held out promise of more high-grade ore. Whitham decided to develop it. However, they could only get the ore to the mill through a separate aerial tram. Such a tram would require a 3,000-foot long span (6000 feet of 7/8-inch cable) with several intermediate towers to reach the mill. Whitham considered it too expensive. He decided instead to drive a 600-foot long drift from the 250 level to the Nugget Tunnel site. Since the Nugget Tunnel was 200 feet higher in elevation than the 250 drift, he reasoned the new drift would go under and on the north side of the Nugget Tunnel and intersect the down dipping gold bearing fault. The company could then mine ore by pulling it down (stoping) and tram it back to the 250 where it would be lowered in the usual way to the mill. The plan sounded good.

Whitham called the new drift the "Nugget Cross-Cut". He started it inside the 250 drift about 200 feet in from the portal. He had not hired an engineer to replace Holdsworth, who was in the Philippines after graduation from the University of Washington. In Holdsworth's absence, Whitham picked the site inside the 250 drift where the Nugget Cross-Cut would start. Because large deposits of magnetite are common at Nabesna, a compass is practically useless in the mine and if used will give incorrect readings. Instead of pointing north the compass needle will point to the nearest magnetite deposit. Unfortunately, Whitham seems to have used a magnetic compass to set the course for the Nugget Cross-Cut.

While Whitham's miners drove the Cross-Cut, other miners were producing ore for milling at a number of other locations in the mine. The Cross-Cut had a relatively low priority compared to income producing work. Only one shift per day worked in the Cross-Cut, and when miners were needed elsewhere work in the Cross-Cut languished. There was no continuity as to the miners who did the work there, either. Often miners other than those who were in the middle of the job in the Cross-Cut resumed the work there. Nevertheless, eventually it reached the 600-foot target distance. But no gold mineralized fault! The miners drove another hundred feet with no better results. Doubt began to dawn. Had they driven the Cross- Cut in the right direction?

Whitham flew in an engineer from Fairbanks. In a few days Whitham learned the bad news. The Nugget Cross-Cut supposed to have gone on the north side of the Nugget Tunnel was instead on the south side! Calculations indicated they had likely missed the intersection of the down dipping gold mineralized fault by several hundred feet.

The mine's diamond core drill went to work and probed northward to try and locate the fault. The drill equipment could reach no further than 250 feet when drilling a flat or horizontal hole and the effort came up empty.

Whitham then had the Cross-Cut turned south. After 500 feet it intersected the down dip section of the 10-foot wide vein the early stamp mill operators had worked on the surface. Whitham and others believed the vein was the southwest extension of the Tower Knob fault system found near the 650 portal. He took some thirty tons from this location and hand-trammed it in three-quarter ton cars 1,200 feet back through the Cross-Cut to the 250 portal. The ore graded two to three ounces gold per ton. That was good ore but a number of problems made development impractical at the time. They portalled out the Cross-Cut on the mountain slope and closed it.

The portal is several hundred feet up the steep slope from the end of Stamp Mill Road. In the summer of 2000 a rockslide partly blocked it. Stamp Mill Road was not built until 1983.

During mining operations ten or so miners worked underground on two

Passenger going up on tram. Loaded bucket coming down.

The aerial tram worked on gravity. The buckets could carry 600 pounds. When the upper bucket was loaded and the brake released it came down one cable and pulled the other bucket up the second cable. Persons wanting to go up climbed to the top of the tower and sat in the bucket. When the operator at the upper terminal released the brake the loaded bucket shot down and the passenger bucket shot up. Sam Gamblin, an old-time miner recalled it was the "most terrifying ride one could imagine."

shifts. The shifts were nine to ten hours a day, seven days a week. The miners were paid $5 per shift plus board and room. The bosses received a dollar more. Christmas and the 4th of July were the only holidays. During this same period miners at Kennecott were paid $4.50 a shift and board and room. Whitham believed he had to pay fifty cents more a day than paid elsewhere. Such are low wages compared to the present, but one must remember this was in the Great Depression. In the states millions were out of work. Five dollars and board and room was a good wage at that time and there was no shortage of men willing to work for it.

Miners rode the tram to the mine portals. Those wanting to "go up" climbed to the top of the lower ore bunker above the mill and sat in the swaying tram bucket. It was only large enough to hold one person if he squatted down or sat on the bucket's edge. The operator at the upper terminal loaded the second bucket with 600 pounds of ore. When the lower operator signaled by the mine's phone system that all was ready,[66] the upper operator threw the

Looking north past Cabin Creek to White Mountain. Photo taken about 1980 before Stamp Mill Road was built. 1. Mill building and town. 2. Tower Knob and 650 portal. 3. 250 portal. 4. 100 portal and site of Bear Vein discovery and the present "Glory Hole." 5. Portal of Nugget Tunnel drift. 6. Sheep Cave. 7. Portal of Nugget Cross-Cut. 8. Stamp Mill Gulch. 9. End of Stamp Mill Road built in 1983. 10. Discovery Gulch. 11. Location of turn-of-the-century Stamp Mill. 12. 1898 White Mountain Cabins. Dash line outlines underground ore body.

tram brake. As the loaded bucket shot down the track cable, so too, did the bucket and miner shoot up the other track cable. It came swinging upward above the trees at break-neck speed and nearing the upper terminal skimmed only feet above the jagged cliff face. Sam Gamblin, a well-known prospector and miner recalled, "The most terrifying ride one could imagine." He told how some new hires simply got off, walked back down the steep trail along the cliff and quit.

Sam Gamblin was one of many interesting characters that worked at Nabesna. He was a prospector and miner, and because he'd been over much of Alaska he'd acquired the moniker "Rambling Sam Gamblin." Time and again it's said he told Whitham and others about 'invisible gold' in the limestone near Discovery and Stamp Mill Gulch where those early miners first discovered placer gold leading to the stamp mill operation. He meant the gold could not be seen in the limestone but assay showed it was there. Whitham called it 'bunk.'[67] However, during the 1990's the USGS heavy metal research at Nabesna seems to support at least some of Sam Gamblin's idea of "invisible gold" when it reported that at five sample sites in the unmined areas that "gold is the most widespread anomalous element."[68]

The mine has four to five miles of underground workings on six levels. If the present day viewer had X-ray eyes, he could see underground workings as tall as a 60-story building that cover more than a city block.

In 1939, Whitham hired Ira B. Joralemon, an internationally known mining consultant from San Francisco.[69] Joralemon spent several weeks examining the mine and made a number of recommendations including continuing work on the Nugget Tunnel and the Bear Vein Fault system and further exploring the Tower Knob Fault system. The placer gold in Discovery and Stamp Mill Gulch, he believed, was from a local source and should be explored. Aware of the underground survey mistakes, he gently admonished, "To guide this work (as recommended in his report) and to keep geologic maps up to date, it is advisable to employ a well-trained young engineer and geologist." He concluded, "It seems likely that with the development ratio kept at a safe level, the prosperous life of the Nabesna may continue for many years."

---

[56] Members of the ARC started calling the trail from the Richardson Highway to Slana the "Abercrombie Trail or Road." The name never stuck. Instead, it had no official name and many called it "Nabesna Road" all the way from the Richardson Highway 104 miles to the mine. Only after the U.S. Army during World War II pushed the road from the Slana Roadhouse through the mountains to the Alaska Highway at what became Tok did the road from the Richardson Highway become called the "Slana-Tok Cutoff." After Statehood the "cutoff" became the Glenn Highway. Wooden Mileposts along Nabesna Road were replaced with metal ones. The last wooden Milepost at the entrance to the old town read "104" meaning that was the distance to the Richardson Highway.
[57] Letter of June 24, 1930, from W.L. Newell Mechanical Engineer to Whitham.
[58] The 100 level was portalled 100 feet vertically below the Bear Vein discovery outcrop and each level portalled thereafter was numbered according to its depth vertically below the Bear Vein outcrop.
[59] Pilgrim, Earl R, 1975, "The Treadwell Mines," *The Alaska Journal, Quarterly*. Vol. 7 No. 4.
[60] For comparison, the largest gold mine in Alaska in 2002, was Fort Knox Mine near Fairbanks and mill-run ore is said to grade something under one tenth of an ounce gold per ton.

[61] Phil R. Holdsworth started working at Nabesna Mine in 1931 as the mine assayer. Later he became the mill superintendent and did much of the underground and surface engineering. His last year at the mine was 1936 and graduating from University Washington he was hired by Mindanao Mother Lode mine in the Philippines, one of the largest gold mine in the Far East. When the Japanese invaded the islands in 1941 he was commissioned in the U.S. Army. He ordered pumps turned off and the mine flooded to prevent the Japanese use of it. He fought with the army and resistance forces. Captured by the Japanese army he and his wife Peggy became prisoners of war in Manila. After the war he returned to Alaska and after a short time in private practice he became Commissioner of Mines for the Territory of Alaska and upon Statehood became the first Commissioner of the State of Alaska Department of Natural Resources. Under his tenure and with advise of Tom Marshall petroleum geologist who as a young infantryman fought his way from Normandy through the hedgerows and that of Charles Herbert a well known and respected mining engineer, a naval officer during World War II, a former Deputy Commissioner, to select the land on the North Slope that became the great Prudhoe Bay oil field. Neither seems to have received the credit they deserve. Those selected lands came under State control.

[62] Pilgrim, Earl R. *Cooperation Between the Territory of Alaska and the U.S. Making Mining Investigation 1931.* Pilgrim was a mining engineer from the Univ. of Washington, a lieutenant in World War I, and in 1922 became the first professor of mining and metallurgy at the Alaska Agricultural College and School of Mines, Fairbanks. He later owned the Stampede Mine in the Mount McKinley district.

[63] Baldwin, U.S. Mineral Surveyor conducted the mineral survey of USMS 1591 in 1932 upon which the U.S. Patent of Nabesna Mine was conveyed. According to others Baldwin probably surveyed the McCarthy townsite.

[64] Baldwin, Asa, 1930, *Report on White Mountain Claims* private report to shareholders.

[65] Janson. L.E., 1975, *The Copper Spike*, Anchorage, Alaska. The Great Depression had caused copper prices to plunge to low levels and Kennecott considered the winter operations were unprofitable.

[66] Whitham had GE battery powered industrial phones in heavy metal cases installed at each tram terminal and in the mill, the assay office, the mess hall and the office in his house. Only the phone in the mine office was connected to the outside ACS phone line.

[67] In the 1950s John Livermore, Newmont geologist and Ralph Roberts, USGS would separately share credit for bringing "invisible gold" to the forefront and it would lead to the "Carlin Trend" in Nevada, an area perhaps second only in gold resource to the Witwatersrand of South Africa. They recognized gold in limestone in parts of Nevada was so small it could not be seen and was overlooked for over a hundred years by prospectors and geologist. An article describing John Livermore and Roberts and others during the Fifties and Sixties in Nevada is by John Seabrock, *Reporter at Large*, New Yorker, April 1989.

[68] Eppinger, R. G., et al, 1995, *Geochemical Data for Environmental Studies at Nabesna and Kennecott, Alaska: Water, Leachates, Stream-Sediments, Heavy-Mineral-Concentrates, and Rock.' USGS., 1st Draft Version,* Open File, USGS, Denver, CO.

[69] Ira B. Joralemon, *Notes on Geology, Nabesna Mining Corporation, Nabesna Dist., Alaska, 1939.* Private report.

CHAPTER 14

Milling Ore and Recovery of Gold

In the Thirties the largest building north of the Wrangell Mountains and south of Fairbanks was the mill at Nabesna Mine. Built on a slope, its four levels had only one purpose, to recover gold. It did that. Out of that building came some five tons of gold, twice that tonnage of silver and many more tons of copper concentrate. It's a rambling building braced by 20 foot long timbers cut at Twin Lakes 15 miles away and hauled by teams of horses to the mill. The men who built it had no modern crane. They hoisted those beams with their strong backs. Today, OSHA, MESA and a host of other government agencies would have shut the work down if it were done like these men did it, they claiming all sorts of human tragedies would happen if they were not there to regulate and protect. No tragedy occurred in the building of the mill. That sort of anecdotal proof that people are smart enough to take care of themselves doesn't seem to deter the regulators. That was a time when men and muscle built things, and the men never thought much about it.

The mill worked in shifts, 24 hours a day, seven days a week. The miners working in the mountain above the mill trammed ore down from the mine portals. From there the mill hands loaded ore out a chute and into a rail car and then dumped the ore into a jaw crusher. The crusher reduced the mine-run ore to three-quarter inch size. The crushed ore then dropped to a bin where an automatic scope fed the ore to the 50-ton ball mill. Loaded with two to three tons of 6-inch steel balls[70] the mill slowly rotated at seven-RPM and the scope continuously fed it. Mill operators added water to insure a slurry of a roughly constant consistency. The cascading balls crushed the ore to 30 mesh, or about the size of table salt. Steel grating at the discharge end of the ball mill allowed only the slurry to flow out but not the steel balls or coarse ore. The slurry entered several rake classifiers, which returned

Over the years the mill building was expanded as production increased and new and different equipment was installed.

oversize to the ball mill for regrinding.

During the early years of the mine's operation was the free-milling period when most of the gold came from high-grade gold bearing quartz ore. The gold in this period was not combined with pyrite and was simply raw gold. This fact dictated a relatively simple separating process.

The slurry then flowed onto a number of concentrating tables where gravity separated the heavy gold from the lighter quartz. The tables were 17 feet long and six feet wide with narrow riffles tacked to their surfaces. These large tables were tilted so when slurry flowed onto the top edge, the gold, being seven times heavier than quartz, hung behind the upper riffles. This was accomplished by a continuous back and forth jerking motion of the tables that caused the gold to sink in the lighter slurry, like heavy marbles in a mud pie. The back and forth shaking motion caused the raw gold behind the upper riffles to slide laterally to the ends of the table where a designated crewman collected it. This gravity system did unavoidably wash quartz particles embedded with small grains or films of gold over the upper rifles and off the middle end of the table. This material bore the name "middlings" because it was said to be collected midway between raw gold and waste quartz that became tailings. The mine saved the "middlings" as a high-grade concentrate. Finally, the system washed the lightest material called gangue to the lowest side of the table and into the tailings. Today one finds this tailing material on the surface of the ground just down slope from the mill. It is nothing more noxious than the mountain itself.

During the operating years the mine carefully confined the tailings by a tailing dam down slope from the mill. The mill was not totally efficient and research in the 1980s show the tailings grade about a quarter of an ounce of gold per ton and more than that in silver.[71] The gold per ton in the mill tails exceeds the grade of ore mined at many of the producing gold mines in Nevada. The Nabesna Mill tailings are one of the few places in Alaska where a pan of material is certain to contain gold values.

During 1931 through 1932 the concentrating tables sometimes recovered less than 50% of gold that was provably in the ore. This meant, in some cases, the mine was losing to the tailings several ounces of gold from every ton of milled ore crossing the tables. The loss was caused when gravity worked against gold recovery by washing tiny particles of gold attached to larger quartz fragments off the table and into the tailing. Nabesna Mining Company could tolerate the loss so long as it was mining rich ore and the tailings were saved for re-milling. Whitham and his crew did re-mill some of the tailings that were particularly high grade.

From the main mill floor of the mill a person could walk to the railing and look down to the second floor and the concentrating tables. The story goes a new hire leaned on the rail and stared down at what was accumulating behind the upper riffles on each table and asked the mill foremen, "Why is all that yellow mustard seed hanging up on those tables?" The foremen told Whitham, "Fire him." Anyone, particularly anyone working in a gold mine, would have known the yellow "mustard seed" was raw gold heaping-up behind the upper riffles. The foreman believed the man's apparent ignorance was a sham and figured he'd hired-on to "high-grade," to steal gold. This was a time when employees were truly "at will" employees.

On those tables, raw gold accumulated every day. Certainly some employees "high-graded" some gold. But at the height of the Great Depression men wanted work and so were not generally willing to risk losing the work they had. Some had walked 104 miles[72] from the Richardson Highway to seek a job at the mine. Consequently, the mining company never reported any major theft of gold. This fact is testimony more to the caliber of men and women in those hard times than any security measures, which were minimal.

The assay office and laboratory were a frame building next to the mill. The laboratory, considering its time and location, was as complete as that

1932 photo of concentrating tables on lower level of mill building. The tables produced raw gold and a highgrade concentrate.

of a small college laboratory. There were two sets of analytical scales with one enclosed in its own dust free cubicle. One furnace was a diesel fired muffle type where flame was applied to and enveloped the outside of a refractory chamber wherein melting took place. The other three types of furnaces were bench-mounted electric furnaces. These were lined with refractory material. An electric current passed through the crushed ore, or charge as it was called, to be melted. The electric furnaces were larger in capacity than the muffle furnace and produced a gold melt of greater fineness. The technicians used the muffle furnace for assay work but used the electric ones to melt and produce bullion. The bullion produced by the assay laboratory seldom exceeded 900 fine. One thousand fine is pure gold. That meant the bullion produced contained about 100 parts per thousand of some other metal, chiefly silver.[73]

Once the mill crew collected the gold, the laboratory technicians melted the raw gold from the tables and cast it into bullion bricks. The middling concentrates ended up in sacks.

Then came the transportation of this precious cargo. A trusted employee, Holdsworth, and sometimes Whitham himself carried the bullion and concentrates worth tens of thousands of dollars by pack train along a trail several miles long to Nabesna landing field alongside Nabesna River. There several now famous Alaska bush pilots flew the cargo of bullion and gold concentrates out to Copper Center or Valdez. These pilots included Harold Gillam and Bob Reeve. From these points agents of the mine shipped the bullion and concentrates to Seattle.

Just by happenstance Whitham started up his mine at a propitious time. Events in 1933 gave the gold industry including Nabesna a breath of fresh air. Congress, because of the banking crisis during the Great Depression, allowed the President to devalue the dollar up to 50%. President Franklin D. Roosevelt finally devalued it by 40%. Congress also required all gold certificates to be redeemed. Finally, Congress, by law, required

Before Nabesna Road reached the mine, gold bullion and concentrate were hauled by packhorses six miles to Nabesna Landing Field and flown out by Bob Reeve, Harold Gillam and other well know bush pilots. Man on the saddle horse is Phil Holdsworth.

that all gold produced by mines and refined be sold to the United States Mint. In doing so Congress fixed the price of gold at $35 per ounce. That was nearly a 75% increase from the price of bullion of $20 per ounce in the year that Whitham started the mine.

About 1934 Phil Holdsworth, foreseeing gold of the lower grade pyrite ore could not be economically saved by tabling, began testing flotation as a method of recovering gold in the mill. When flotation showed promise of recovering gold otherwise lost to the tailings, Whitham purchased the mine's first flotation units and freighted them to Nabesna.

Separation of the gold from the ore by flotation required a finer grind from the mill. Whitham installed a new ball mill that reduced the ore to the size of talcum powder (-60 mesh). As before, the mill workers added water to create a slurry in the ball mill. From the ball mill the system piped the slurry to classifiers. From there the slurry flowed to concentrating tables to recover any raw gold in the old way. All products from the tables, except the raw gold, went to a large flotation cell called a Denver Sub-A. There the mine recovered more raw gold, some of that which had escaped over the tables. From the Denver Sub -A, the system directed the slurry to a bank of flotation cells. Each cell was about the size of an average home refrigerator. The flotation operators added a few drops of pine oil by way of a reagent feeder. Each cell was equipped with an impeller, basically a spinning propeller at the bottom of the tank. When the system forced compressed air into the spinning slurry, bubbles resulted. By design, the rising bubbles became coated with pine oil. Tiny, even microscopic size gold particles adhered to the rising pine oil coated bubbles. The mill operators called the bubbles "froth" when they surfaced. A flapper on each tank automatically skimmed the gold bearing froth off and collected it. Certain cells collected pyrite still combined with gold. The mine saved this as a concentrate that the men also called "middlings."

The first flotation cells purchased by the mine lost nearly 30% of the available gold. Shortly, the mine replaced them with more modern, by 1935 standards, cells.

The ore the miners sent down to the mill continued to be of both kinds: quartz ore with raw gold and skarn ore with gold-pyrite. However, as mining went deeper and away from the rich Bear Vein fault, the volume of lower grade gold-pyrite ore increased in proportion to the total. It was more difficult to save the gold-pyrite by flotation than it was to save the raw gold on the tables. Loss of gold from the flotation circuit continued until the loss sometimes exceeded 20 percent. If gold could be separated from the pyrite at the mine, the high loss of gold could be reduced and the cost of shipping worthless pyrite in the concentrate to the smelter could be saved. The Mine would also greatly reduce its smelter charges. The smelter penalized the mine for high pyrite content of the concentrates.

Holdsworth had built the assay laboratory at Nabesna Mine into a highly efficient one. By mid -1930 he was experimenting with cyanidation at the mine. And during the winter semester at the University of Washington he demonstrated cyanidation could economically free the gold from the pyrite and produce a gold-silver cake that could be melted in the assay lab to a valuable product called a "dore;" a mixture of gold and silver. Producing a "dore" at the mill would eliminate the expense of shipping tons of worthless pyrite to the Tacoma Smelter.

Cyanidation can separate, in this case, gold and silver from pyrite by a chemical process. This process requires large volumes of water, half a dozen large tanks, thickeners, agitators, filters and pumps. These tanks remain in the mill and appear like hot tubs for giants. Cyanidation at Nabesna used a very weak solution of one pound of cyanide per ton of water. Temperature of the solutions is critical as is the acidity and alkalinity. In short, the operator has to know what he's doing. The cyanide used at Nabesna was manufactured in Switzerland and came in wooden crates holding 50 pounds of snow-white one-pound blocks. It cost $18 per crate, freight included.

The "how" of the process is complicated but simplistically can be explained this way: When a crystal of pyrite with included gold and silver is submerged in a weak cyanide solution the cyanide will dissolve the gold and silver and leave the pyrite more or less unchanged. The operators then recover the dissolved gold and silver from the cyanide solution by pumping the solution into wooden tanks where zinc shavings were suspended in movable racks above the bottom. The cyanide forms a compound with the zinc and the gold and silver is "precipitated." The precipitate is called a cake and looked altogether like gray mud at the bottom of the zinc tanks. The cake is washed from the tanks and taken to the lab where it is dried and then melted in the furnaces. The final furnace product is a mixture of gold and silver together with some impurities, the so-called "dore." Cyanide is quickly destroyed by sunlight. Testing conducted a half-century later at Nabesna in the 1990s by the Bureau of Mines, Environmental Protection Agency, National Park Service and the U.S. Geological Survey revealed no danger of cyanide. Often overlooked is that a number of common plants contain a natural cyanide compound.

Holdsworth designed and oversaw construction of the cyanide plant. Under his management it worked efficiently at the mine. However, because the process is complicated and tolerances are critical, only experienced and careful operators manage to recover gold efficiently by the process. As already mentioned Holdsworth graduated from the University of Washington in 1937 and left for the Philippines for a job offered him there in one of the Asia's largest gold mines.

Afterwards, at Nabesna, matters didn't turn out well for the new cyanide process. Within a year Whitham discontinued the cyanidation plant and returned to flotation. He refined the process and thereafter lost about 10 percent of the gold-pyrite to the tailings. He was willing to accept that loss believing the gold could be recovered in the future by reprocessing the tailings.

The mill continued to produce two products: a gold-silver bullion from the assay furnaces and a high-grade concentrate of gold bearing pyrite. The lab technicians weighed each bar of bullion at the mine then shipped it directly to the United States Assay Office in Seattle. The Seattle Assay Office officially weighed the bars. If there was a difference between the two weights, the Seattle office notified the mine. Small differences of hundredths of an ounce did occur but only rarely. Never major differences, except once. In March 1938, after Holdsworth had left, the United States Assayer in Seattle wired the mine that the mine's weigh of bullion received was higher by 20 ounces than the official weight? Whitham perceived this as a real slap at the mine. He wired back immediately the mine would accept the U.S. Assayer weight. No record exists of the language, and it could be pretty salty, Whitham directed at the person responsible for the embarrassment.

As for the concentrates, the laboratory crew placed about one hundred pounds in a canvas bag and then put that bag in a tightly woven burlap bag imported from India. Thusly, the concentrates were shipped. Before shipping the crew took a representative sample from each bag, assayed it and, knowing the weight of concentrate in each bag, calculated the gold-silver value of each bag. A Seattle insurer insured this value. The mine shipped the concentrates directly to A.L. Glover, Inc., assayers in Seattle. There, A.L. Glover assayed the concentrates a second time. Glover then forwarded the concentrate to Tacoma Smelter owned by Kennecott. Tacoma Smelter assayed the concentrate again before refining it to gold and silver bullion to be sold to the US Mint.[74] All gold shipments were insured from Valdez to Tacoma but not from the mine to Valdez.

Whitham built his mill in stages and enlarged it as the milling process changed. At peak production the mill could process 100 tons per 24 hours. Overhead belts turned by a single electric motor in turn powered by a diesel-powered electric generator, drove the mill equipment in the early years. Eventually Whitham upgraded the power plant and replaced the old belt drive system with electric motors on individual units. The thickets of insulators in the mill building speak mutely of miles of insulated wire connecting the various units throughout the mill.

Form 42—10M—5-37—CDT

Copper River & Northwestern Railway

| Number | ## 7 | Time Filed | | Check | 44 Collect N.L. |

Send the following message, subject to the terms and conditions on the reverse side of this blank, which are agreed to by the undersigned sender.

Seattle Washington
March 14 th 1938

Nabesna Mining Corp
Chitina, Alaska.

RETEL FEBRUARY TWENTY FIFTH BAR NUMBER FORTY WEIGHS EIGHTY FOUR DESIMAL SIXTY FIVE TROY OUNCES INSTEAD OF ONE HUNDRED FOUR DESIMAL THIRTY ONE AS CLAIMED BY YOU PACKAGE ARRIVED IN APPARENT GOOD ORDER SHIPMENT BEING HELD PENDING YOUR ACCEPTANCE OF OUR WEIGHTS

G L SWARVA
ASSAYOR IN CHARGE
A.F.***C.F.P. 4.46 P.M.

Rate 50 wds   2.00
Tax   .10
                     2.10 Collect
Charge N M C and will collect on copy.

Form 42—10M—5-37—CDT

Copper River & Northwestern Railway

| Number | ##.1 | Time Filed | | Check | 10 | PD BLK |

Send the following message, subject to the terms and conditions on the reverse side of this blank, which are agreed to by the undersigned sender.

CHITINA ALASKA MARCH 15 1938

G L SWARVA
ASSAYER IN CHARGE
US ASSAY OFFICE
SEATTLE WASHINGTON

REYOURTEL FOURTEENTH WILL ACCEPT YOUR WEIGHT OUR BAR NUMBER FORTY

NABESNA MINING CORP

C.F.P.--- 27 905 P.M.
Rate 10 wds   2.00
Tax   .10
                2.10 Paid

Charge N M C

CHG NMC

Law required gold produced by Nabesna Mine be sold to the U.S. Mint for $20 per ounce before 1934 and $35 per ounce afterwards. U.S. Mint in this case checked the mine's weight and found an error. The mine accepted the Mint's weight.

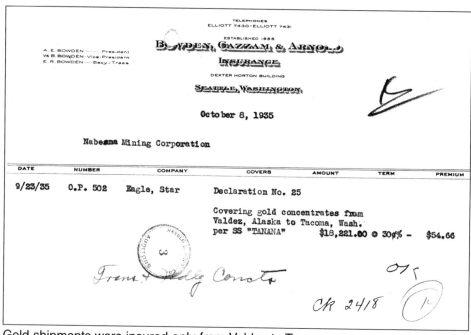

Gold shipments were insured only from Valdez to Tacoma.

Nabesna power plant located in the mill building furnished electricity for the mill, mine and town.

[70] Walking around the old town, one gets the impression that it was once a training camp for shot putters.

[71] Cleveland, G,. 1981, *Final Report on Nabesna Mill Tailings*, WGM, INC. Mining and Geological Consultants. Private report. pp 52.

[72] Alongside Nabesna Road at the entrance of the mine town was a Alaska Road Commission wooden Mile Post sign that read "104," meaning it was 104 miles from the Richardson Highway. Soon after the area was included in the Wrangell-St. Elias National Park and Preserve in 1980 the sign disappeared.

[73] Carat or karat is most often used in the jewelry business whereas fineness is the term used in the mining industry. Twenty-four carat is pure gold and 1000 fine is pure gold.

[74] Nabesna Gold Mine by law had to sell the gold produced after 1934 to the federal government for $35 per ounce. That gold is in Fort Knox. In 2002 that gold was worth more than $300 per ounce. The federal government has therefore made a profit of several millions of dollars from the Nabesna Gold Mine.

CHAPTER 15

Nabesna Gold Mine Town

Nabesna in its time was one of the most isolated hardrock mines in the Territory of Alaska, which mandated a large measure of self-sufficiency. Whitham had to build a town. By 1932 Whitham directed that the tent camp be replaced by log and frame buildings. The mine's on-site sawmill furnished the lumber.

Whitham saw to it a medical dispensary was in place. Considering the town's remoteness, it was well equipped. For serious cases, Nabesna Mine had an arrangement with Kennecott Copper Corporation such that Nabesna could send injured or sick miners there to its fully equipped hospital. The copper mine was 70 miles by airplane across the Wrangell Mountains. Nabesna Mine also flew the injured to St. Joseph Hospital 250 miles away in Fairbanks. The mine owned and operated a fully stocked retail store of groceries and dry goods. Employees, prospectors, government surveyors and anyone else who found himself or herself in that remote area, could buy there. One evidence of the wide span of ages of the general store's customers is a wholesaler's invoice whereon is listed boxes and boxes of snuff, and cartons and cartons of Baby Ruth, Milky Ways, and Snickers. The company sold all items at cost.

Carl Whitham occupied his own house. It was an attractive log structure, relatively small with two identically sized rooms. The rear room served as the company office. Whitham's house is still up there behind the mess hall. The mill superintendent and the mine bosses had their separate cabins, too. Single men stayed in bunk houses of which there were a number. Some bunkhouses were two story affairs. Many structures have been burned or scavenged to nothingness over the years. The mine provided married workers cabins off the patented property. Visitors had access to guesthouses. The town had its own laundry facility, in the mill. The miners could clean their own clothes or pay the laundry for the service.

Following several fires that destroyed or seriously damaged bunkhouses and other buildings, Whitham ordered installation of a large boiler in a separate building. Its tall black stack remains a prominent feature in the town. From the time the mine installed the boiler most of the town and mill was steam heated. The many iron radiators found in the old town are a reminder of this interesting feature. Those cast iron radiators are almost immovable and remind one of the staggering job of transporting all this equipment to this remote place over trails from the coast of Alaska. To get the steam to the structures required piping. One finds hundreds of feet of iron pipe underground in the town, testimony to a heating system that was as efficient as it was amazing. In effect, it was central heating for a whole town. That was just the beginning.

The power plant in the mill supplied electricity to the town. The mess hall could feed at least 30 at a sitting and had its own bakery. It was open 24 hours every day to all that passed that way. Early on, Whitham hired local hunters to supply fresh sheep, moose and caribou. Nabesna country was then as it is now, a well-known big game hunting area, and hunters often left fresh meat for the messhall. When Nabesna Road reached the mine, trucks could travel there and the mess hall could purchase fresh beef, pork and chicken from farms near Valdez and Kenney Lake. The mess hall remains standing in the old town. It is the anchor building of the boardwalk, which connects the front row of buildings. The town is nothing if not interesting even now, even though many of the original buildings have been lost.

During the early years the mine closed in October and reopened in March. Part of the reason for closing seasonally was the need for mill water. In the summers of the early years the mill received water by a several thousand foot long wooden flume from Cabin Creek. Part of it is still there in the trees south of town. However, by October it froze. Later the mine put in water wells and solved the freezing problem. Further compounding winter operations' problems after 1932 was the winter shutdown of the Copper River and Northwestern Railway. The alternate route out of Valdez over Thompson Pass on the Richardson Highway was stymied by world record snowfall in the winter. In that pass snow is not measured in feet but in tens of feet. That snow bound Pass always closed the Richardson Highway out of Valdez from October to May.[75] Airplanes during that period became the only transportation in and out of Nabesna. The cost of flying freight was $200 to $300 per ton.

After 1935, when Nabesna Road reached the mine in roughly its present shape, both meanings intended, the mine's cost of transportation diminished substantially. That still did not eliminate the need for bush planes and especially so in winters.

When the mine had a special need for miners, Bob Reeve Airways out of Valdez, or Gillam Airways out of Fairbanks, Lyle Airways out of Copper Center, Cordova Air Service out of Cordova and Pollack Flying Service from Fairbanks flew them in. The charter fare for one man from Cordova to Nabesna was $45, from Valdez $40, and from Fairbanks $140.

In 1900 Congress appropriated funds to provide telegraph service for the Territory. The Alaska Communication Service (ACS) organized within the U.S. Army Signal Corps provided that service. Soldiers strung the wire on pole tripods along the 450-plus mile long Eagle Trail between Valdez on the coast and Eagle City on the Yukon River.[76] Typically the signal corps placed telegraph stations every 50 to 70 miles along the line. Soldiers manned these remote and often isolated stations. Eventually, the corps upgraded the telegraph service to telephone service.

Despite earlier unsuccessful efforts by Whitham to secure telegraph service to the mine, in 1932 he succeeded in inducing ACS to string a telephone line from Lawrence DeWitt's Slana Roadhouse 46 miles to the mine. The Army strung it by hanging the iron wire on insulators attached to pole tripods cut from local spruce trees along Nabesna Mine trail. Since World War II, when ACS discontinued use of the line, that 46 miles of wire has served many a homesteader with clothesline. Occasionally an alert traveler along that road may see bleached tripods in the brush.

Early start of the town. Back of photo reads "Nabesna 1932." Some tents are still in place.

In 1932, Whitham applied to the General Land Office (GLO) for a mineral survey of the company's White Mountain mining claims (after World War II the General Land Office became the Bureau of Land Management). The GLO in due course completed the survey, U. S. Mineral Survey 1591. The company paid for all costs of the survey. President Franklin D. Roosevelt signed and issued the patent to Nabesna Mining Company in 1933. Those patented claims encompass the town and much of White Mountain. This property thereby became privately owned land.

The company expended hundreds of thousands of dollars on the claims, and during the Great Depression hundreds of people earned livings because of the mine and considered that mine a godsend. The right of the mine's owner to be there and to offer those jobs was insured by the mine's fee ownership, the patent rights, of the claims. People who criticize patenting of mined lands might well look at the Nabesna experience for another view. Anyone questioning the morality or wisdom of patent for Nabesna apparently gives little thought to the labor, trials and tribulations overcome by courageous men to build that mine. They might also consider that the mine perhaps would not have been built except with knowledge by its builders that they could secure title rights that could not willy-nilly be taken away by some bureaucrat. Moreover, persons who criticize mining patents tend never to consider the value of the land patented before the mining development.

The present Nabesna Road reached the mine in late 1933. At first it was nothing more than a widened trail. The only bridge across numerous streams that cross the road was across the Slana River. The dozen or more other creeks had to be forded, as many still must. Only trucks freighting equipment and supplies used the early road before heavy snows blocked it in mid winter. Not until 1935 did the road see improvement sufficient to allow for bus or car travel. Even so it took a good part of a day to drive the 46 miles from Slana Roadhouse to the mine. John Kelsey[77] tells that upon graduation from Valdez High School in 1938 and in the summer while attending Stanford University he drove a truck for Alaska Freight Line,[78] hauling supplies in and bullion and concentrates out of Nabesna Mine. "The drive from Valdez to the mine took two days," Kelsey said, "provided we didn't get stuck on Nabesna Road." Today the drive from Valdez to the mine takes about eight hours…provided one doesn't get stuck on Nabesna Road!

Many well-known longtime Alaskans knew Carl Whitham, and when they were young men worked for him at the mine or handling gold concentrate at Valdez or elsewhere. All had a good word for him, including George Sullivan.[79]

Visitors were always arriving at Nabesna and some for unexpected reason. Once it was Bob Reeve.[80] Reeve did a lot of flying for the mine and was a close friend of Whitham. In a book about Reeve, *Glacier Pilot*, he tells how after flying mining supplies to Chisana, on his return to Valdez while flying at 14,000 feet to cross the Wrangell Mountains, his engine quit. The only airfield in the vicinity was the mine's Nabesna landing field but he couldn't make it. He had to land in the steep walled canyon of Jacksina Creek below the mine. It was March. That means it was cold. Snow was three feet deep. After getting the plane anchored down in the frozen ground, Reeve and his passenger set out on foot for Nabesna Mine five miles away. "Wind was blowing like hell", he wrote. They reached the old town after midnight and Reeve pounded on his friend, Carl Whitham's, door. Awakened from a sound sleep, "Carl nonetheless greeted me like an old friend and promptly pulled out a jug of Doc Blalock's mountain dew." In other words, moonshine. The next morning Whitham sent a four-horse team and sled to pull the plane to a safer place than where Reeve had been forced to leave it. Afterwards, Whitham had a dog team carry Reeve 46 miles to Slana Roadhouse. Eventually Reeve repaired his plane and flew it out.[81]

Father Bernard Hubbard, S.J.[82] a Professor of Geology, University Santa Clara, California, stayed at the mine while conducting his research on Nabesna and other glaciers. Nabesna Glacier ends 14 miles south of the mine. It is little wonder the priest found this glacier, perhaps the worlds' longest, fascinating. Whitham furnished Father Hubbard supplies as well as saddle and packhorses on trips to study the glacier. Whitham always reserved a cabin at the mine for his use.

Harold Gillam's open-cockpit "Zenith" on skis at Jack Lake along Nabesna Road 15 miles from the mine. During the winter cars could readily travel Nabesna Road from the mine to Jack Lake.

During the 1930s Father Bernard R. Hubbard, S.J., professor of geology at the University of Santa Clara, was a friend of Whitham and stayed at Nabesna while doing research on Nabesna Glacier. Father Hubbard was well-known in Alaska and often called the Glacier Priest. He was a friend of General George C. Patton and visited him at his headquarters in Europe after World War II.

During the twenties and thirties the Copper River & Northwestern Railroad, in cooperation with the Cordova and Valdez Chamber of Commerce, pushed tourism to the Interior. One could enjoy a first-class round-trip lasting three weeks from Seattle to Cordova-Kennecott-Valdez and back to Seattle for less than $200. By the mid -thirties when Nabesna Road had improved "somewhat" the very adventurous could make the drive to the mine.

Territorial Governor Ernest Gruening was one such adventurous person. He was impressed with the Wrangell Mountains. He was admirably prescient from those pre-World War II days in believing tourism held economic promise for Alaska. On one trip to Nabesna Mine, he stayed several days and was said to be much impressed with the Nabesna country. What he did not like is still an element of getting to that country today. Whitham, always a gracious host simply commented "The governor didn't much care for Nabesna Road."

Another visitor to the mine in its glory days was Dr. Charles E. Bunnell, President of the University of Alaska at Fairbanks. It was then the only university in Alaska. He made the trip with several students of the University's School of Mines. Of several recollections of his visits were his comments of the miners' "outhouses." Each was a doorless shed about 10 feet long by 6 feet wide sitting atop a deep open trench. Over that trench and anchored to the ends of the sheds stretched two long sturdy spruce poles, one to sit on the other to lean back on. At least 10 miners could sit side by side on this arrangement and answer nature's call. An old time miner recalled what the "Judge" said on seeing the outhouse in use[83] (old timers most often called Dr. Bunnell, "Judge") "The crew sat there chattering like a bunch of magpies." Those for women and visitors were more discrete and some even had doors. On visiting the old town, EA Patterson of Fort Worth, Texas enjoyed the view from one such facility so much that she

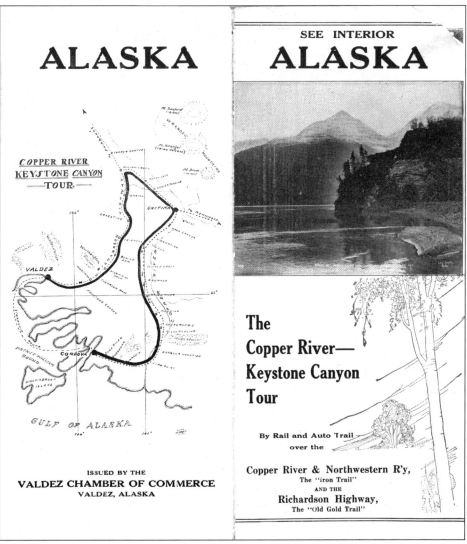

The Copper River and Northwestern Railroad was built to haul copper ore from Kennecott Mines to Cordova, then shipped by Alaska Steamship Company to the Tacoma Smelter. The train always had several passenger cars and when encouraged by the Chamber of Commerce of Cordova and Valdez it entered the tourist business and successfully so. The mines and railroad closed in 1938.

Of the amenities at the old town indoor toilets was not one. Seeing miners sitting 10 at a time, Dr. Bunnell President of University of Alaska dryly remarked, "They sat there chattering like a bunch of magpies." Those for guests were better appointed and offered a grand view of the valley. Some lingered, causing Dr. John Patterson of Fort Worth, Texas to come looking for his wife, EA, and thus the photo opportunity.

would remain there viewing the great expanse of valley and mountains. Consequently, her husband, Dr. Patterson,[84] went looking for her, accounting for this photo opportunity.

Another visitor story goes that two young lady professors from some "Outside" university visiting the mine wished to go underground.[85] An underground mine is to the uninformed a strange and foreboding place. Whitham, knowing this, was reluctant to permit the excursion. However, Whitham, after much coaxing, finally agreed and sent two young miners with the professors.

Taken up the tram, an unnerving beginning, the guides fitted the lady professors with borrowed carbide lamps and miner's caps. With that faint pungent odor of burning acetylene from the carbide lights trailing in their wakes they strode into the 250 portal. Footfalls echoed dully around them as they trooped deeper into the mine. No place on earth is so dark as an underground working deep in the solid rock of a mine. The only light was that fixed to the miner's cap and every movement brought weird shadows dancing along rock walls. Wisping faintly of sulfur, that black hole is like Hell itself might be.

In that total blackness, the brain becomes clouded and the very air reeks of something unseen, as if some evil is lurking there just out of sight. Those dark caverns deep in White Mountain twist and turn. The unwary lose all sense of direction. It soon becomes apparent that escape is wholly at the whim of the guide who carelessly, so it seems, forges ever deeper–perhaps to the point of no return. At a point fear submerges curiosity.

This was likely the mental condition of the two women as they ventured into the depths of the mine. And, as the story goes, the miner's lights of the foursome in one of those dark uninviting drifts highlighted a plank laid across two empty wooden dynamite boxes shoved against the rock wall of the drift. The miner guides thoughtfully suggested they sit and eat their lunch the cook sent along. The story doesn't indicate if the ladies ate. Those lunches were probably by then the only reminders of a wonderful world of light and color they never expected to see again.

Sitting uneasily on the plank one lady asked if anyone had died in the mine. She, no doubt, thought she knew the answer and already thought she was certain to die there. "Yes," said one young miner and explained the corpse had probably been laid-out on the very plank they were sitting on. Nervous shifting caused one of the empty wooden dynamite boxes to twist then break apart. The plank gave way and snapped like a shot. Bodies spilled across the rock floor and carbide lights snuffed out. High-pitched screams split the blackness. And, as old timers like to tell Whitham told how one of the hysterical ladies screamed, "I want outt'a this f——— place!" The two young miners, so he liked to say, were shocked because they didn't believe an educated lady knew such a word–let alone would shout it.

Nabesna post office was re-established in the early Thirties. Whitham served as the first Postmaster of the re-established office. Eventually, a mine employee filled the job. That old building stands up slope from the mess hall just in front of Whitham's cabin. Back then, being the only post office in the entire district, it was, as one might expect, an active place. However, the Post Office was not the only active building.

Recent photo of Nabesna post office building. Established in 1909. D.C. Bud Sargent was the first postmaster. Early on mail was delivered by dogsled in winter and pack horses in the summer.

    Another was called the "Do Drop Inn." The activity there was "different." A lady proprietor, who in turn employed several other ladies, ran that establishment. On one occasion, perhaps due to the rush of business of the holidays, the proprietress had neglected to renew her liquor license for the coming year, 1938. Knowing the winter mail service was slow she was unsure her check would reach the proper authorities by 1 January 1938. So, being a law-abiding lady and wanting no trouble with liquor authorities, she used Whitham's phone and on December 31 called the mine's office in Chitina to have someone there forward a telegraphed message to the Anchorage Clerk of Court asking:

Wherever miners are so are women. Whitham didn't judge this arrangement and had no complaint about it so long as activities did not interfere with work. Notation on this photo is "Jane and her dog sled."

> "Due winter mail service only once month will not leave
> Nabesna till January fifteen advise collect wire whether
> can operate liquor store after today."

The Anchorage Clerk of Court seemed unsympathetic and wired collect,

> "Sorry but cannot operate suggest sending mail to Cordova
> Will be there about February first."

It is not known if the lady sold liquor after midnight of Dec. 31, 1937. However, New Year's Eve in that remote mine town without booze? Not a chance.

The 'Do Drop Inn' moved several times. One location was in what is left of a cabin just northeast of the old town. The last location was alongside Nabesna Road near Skookum Creek. Part of it is still there. Most travelers pass it by without recognizing its interesting heritage, seeing only a few tumbled down, age-weathered logs grown over by brush.

Wherever miners are, so are women and Nabesna was no exception. Whitham didn't judge this arrangement and had no complaint about it so long as activities did not interfere with work. The ladies were young, good looking, ambitious and never hesitated to do other work, and be paid.

One report of the ladies concerns a young Territorial mining engineer (later a prominent mining engineer) making an inspection of the mine. On noticing several ladies he asked Whitham what they did. "That's the damnest question!" Whitham is supposed to have answered.

The young, attractive and enterprising ladies had pretty much the run of the place, except for the assay office. There the young, male laboratory workers melted gold and silver bullion and poured it into bar molds. Whitham explained, "No sense tempting anyone."

As is true of other frontier mines, many persons passed that way, worked for a time, made a "stake," and then traveled on. Often they didn't use their real names. Some didn't curry close friendships and drifted into obscurity. That was the way with some of the ladies, too, who called Nabesna home for a time.

Up slope of the other buildings, hidden in the timber and brush, is a small cabin. It is such that even a glance tells it was built far differently than others. Its roof is laid with discarded and flattened five-gallon buckets that are now rusty. The old house thus gets its name "the tin house". The rest of the cabin is of similar discarded logs and boards. The story of the origin of this cabin goes that a miner fresh from Italy[86] had portalled the 650 and had drifted some 10 feet into the rock when he loaded too much dynamite in the drill holes and shot it. Mill hands heard the explo-

```
Form 42—10M—5-37—CDT        37 wds 1 extra rush
```

Copper River & Northwestern Railway

| | 37 Paid Black |
| Number ## 1   Time Filed   Check   1 extra rush PD BLK |

Send the following message, subject to the terms and conditions on the reverse side of this blank, which are agreed to by the undersigned sender.

CHITINA ALASKA DECEMBER 31 1937

DERICK LANE
CLERK OF COURT
ANCHORAGE ALASKA

APPLICATION AND LICENSE FEE RENEWAL RETAIL LIQUOR LICENSE MAILED
NABESNA DECEMBER TWENTIETH DUE WINTER MAIL SERVICE ONLY ONCE MONTH
WILL NOT LEAVE NABESNA TILL JANUARY FIFTEENTH ADVISE COLLECT WIRE
WHETHER CAN OPERATE LIQUOR STORE AFTER TODAY

                              BERNICE HOLMES
                     C.F.P.--   a 7 1102   A.M.
Rate 10 wds   1.20
27 Extra at .10 ¢ per each wd 2.70
                               3.90
Tax   .20
                               4.10 Paid
                                    Chge W P C
         CHG WPC

```
Form 42—10M—5-37—CDT
```

Copper River & Northwestern Railway

| Number ## 2   Time Filed   Check   16 Collect |

Send the following message, subject to the terms and conditions on the reverse side of this blank, which are agreed to by the undersigned sender.

                   Anchorage Alaska.
                   December 31 st 1937
                   And January 1 st 1938
Bernice Holmes
Chitina

SORRY BUT CANNOT OPERATE SUGGEST SENDING MAIL TO Cordova WILL BE THERE ABOUT
FEBRUARY FIRST

                              DERICK LANE
              A F -- CF.P. 10.30 A.M.

Rate 10 wds   1.20
5 Extra at .10 ¢ per each wd   .50
                               1.70
Tax   .09
                               1.79 Collect
      Charge W P C and will collect on Orignal

The "Do Drop Inn" was run by a lady proprietor and perhaps by the rush of holiday business she had neglected to renew her liquor license for the coming year of 1938. Being a law abiding lady she used Whitham's phone and called the mine's office in Chitina to have them send the above wire to the Anchorage Court. The court turned down her request to operate without a renewed license. One wonders at that remote mine did the "Do Drop Inn" celebrate New Year's Eve 1937 without booze? Not a chance!

The "Do Drop Inn" was at several locations on the mine property. In time it seems Whitham figured it was just too convenient and miners were spending too much time there. The last location was along side Nabesna Road near Skookum Creek. Travelers now pass it by seeing only a few tumbled down logs grown over by brush.

sion a thousand feet down slope as if they were right there. Those persons in the town heard it, too, and saw mine timbers, rock, and dust boil out the portal on the mountain slope. That evening as the miner came into the mess hall someone asked why the thunderous explosion. Arms and hands gesturing wildly and in broken English the new Italian immigrant exclaimed, "She went boom!" Thereafter as the miner strode into the messhall a chorus of cheers met him, "She went boom."

In time it was too much for the miner and Whitham agreed he could build his own cabin and eat there too. Thereby the "tin house" came to be. In time the miner and architect of the "tin house" moved along. After his occupancy the "tin house'" served occasionally as a secluded place of rendezvous, then as a place for a still for moonshine. Finally, it grew vacant, forgotten and hidden in the brush.

Winter sports, except staying warm, were not a popular activity at the mine. But one winter a young miner and student from the Alaska Agricultural College and School of Mines, Fairbanks, had a different idea. He worked through Christmas at the mine, and, before quitting to return for the second semester at the college, he undertook something for which he had apparently carefully laid plans.[87] Seems he was a ski enthusiast and a skiing friend of Ivar Skarland at the College.[88]

This cabin is called the "tin house" for obvious reason. It is in the brush back of town. It was built by an Italian miner named Bonito to avoid teasing by other miners. In time the miner left and the "tin house" was used for other purposes, including a moonshine still, and in time it was forgotten.

One of life's beautiful mysteries is how different people will look at the same thing and see very different things. White Mountain, rich as it might be in gold, to the young miner-student represented something even more interesting than a repository of gold. He saw it as an untried downhill ski challenge. He used the tram to the 250, and then packing skis, he climbed the underground manway up to the 100 level. There he crossed a rocky ridge to the sweeping, 3,000 feet long snow-packed talus slope, and down he came. Bruised and scratched by brush in the lower area he declared success, left the mine and was never heard from again. No record exists of anyone else trying the slope.

While the skier managed his run without significant injury, the mine had its accidents. Whitham seems to have contributed his share of them. There was the time Whitham fell down a 100-foot manway in the mine. Three miners had to hoist him up to the 250 Level where they tied him in the tram bucket and lowered him down, no doubt more slowly than was customary. He had a broken arm, a broken leg, some ribs were broken and in the words of the mine's dispensary record, "He was in pretty bad shape and moaning quite a bit." He needed more treatment than was available at the mine dispensary, so his people sent out word they needed help. His friend Bob Reeve was in Fairbanks but his plane's engine was being worked on and couldn't help. The mine then got word to Harold Gillam who was in Cordova. In a blinding snowstorm Gilliam took off and in that storm flew all the way to Nabesna. He picked up Whitham and flew him to St. Joseph Hospital in Fairbanks.

Inze Bronniche, wife of Fred Bronniche, was a journalism graduate of the University of Minnesota. In the mid-thirties she and Fred came to Slana. Being young they planned to rough it and go placer mining on their own. They did for a while. Inze knew Nabesna Mine, Whitham and practically everyone else who worked there. After World War II she published some stories about that part of Alaska. She could tell many first-hand accounts and had access to many second-hand ones. One of her favorites had to do with another of Whitham's accidents. "Carl," so she related, "had the habit of going up the tram to the mine after supper. Harry Boyden was the lower tram operator." Boyden was an Englishman, a long time Alaskan, and is said to be the first to bring a packtrain across the Wrangell Mountains.[89] As Harry Boyden related to Inze, "Carl was in his cups." He climbed the lower tram tower, got in the bucket and told Boyden to call the upper operator to release the brake. The bucket arrived at the upper terminal, and when it did the upper operator, expecting to see his boss momentarily, called down to Boyden and asked why Whitham had not come up. Telling the story, Inze liked to mimic Boyden's proper English accent, "Well now, Carl must be in between." A search found Whitham lying in the brush under the tramway cables between the lower terminal and the 650 terminal. He had fallen out of the tram bucket. A leg was broken. When asked, "how come the accidents" an old timer simply said, "Carl wasn't the biggest man around but he was just plain tough and he liked a drink or so."

Whitham did like a drink or so, and according to a well-known lawyer and life-long Alaskan, on one particular steamship trip to Seattle, Whitham was several "sheets-to-the-wind" and apparently unacceptably boisterous. The bartender asked him to leave the ship's saloon. He agreed but before doing so offered to buy the "house" a drink. When the bartender declined, Whitham replied, " Well then I'll just come across and do it myself." With broken tables and chairs littering the saloon floor, the ship's crewmen wrestled him away. The captain had him thrown in the brig for the rest of the trip.

The mine employed 70 or so persons, mostly men. Nabesna Mine was the largest employer north of the Wrangell Mountains and south of Fairbanks in the depression-era. Before Nabesna Road was built, job seekers got there by horseback, dog sled or simply walked. To insure their safety from storms, Whitham had several shelter cabins built along the trail between the Slana Roadhouse and the mine. After Nabesna Road reached the mine, those hustling for a job could, if lucky, catch a ride on a truck hauling supplies in and concentrates out.

During the thirties a number of individuals built homes and businesses on and off the mine property. Almost all are gone, but a two-story house remains. It is now part of the historic and well-known Devil's Mountain Lodge with its own airfield alongside Nabesna Road a few miles short of Nabesna Mine.

During the summer months Dr. Fred Moffit, USGS geologist, a long time friend of Whitham, mapped the Eastern Alaska Range, a project begun in 1905. In the thirties he supplied out of the mine. In 1940 he completed his monumental work "Geology of the Eastern Alaska Range," He'd done it all on foot and horseback. In 1940, with Moffit was a young assistant geologist, Russell G. Wayland, a recent PhD graduate of the University of Minnesota. Moffit had Wayland spend some time at Nabesna Mine.

Photo of Carl F. Whitham in the spring of 1939. He is standing beside the mill and using a cane recovering from a fall in the mine.

Wayland credits Ira B. Joralemon for the mine maps he used, but most of those maps were actually based on Holdsworth's surveys. One might argue with some of Dr. Wayland's conclusions but not his mineralogy or petrography.[90]

Many mining companies then, and some today, consider their stockholders should expect to take their profits by buying shares low and selling high and should not expect dividends. Obviously, mining shares were not then and perhaps are not now the choice of working families and retirees. Whitham thought differently. He believed those who bought shares of Nabesna Mining Corporation should receive income when it made a profit. In its third year of operation, according to public records, Nabesna Mine paid dividends equal to its total capitalization. No other mine in Alaska equaled that. Nabesna Mining Corporation never borrowed money and paid all expenses from earnings.

The historic two-story house was built by the mine. Later it served as the mine's watchmen's house. Along with other historic buildings, it is alongside Nabesna Road about 3 miles from Nabesna Mine. It and other historic buildings are now part of the popular Devils Mountain Lodge. White Mountain looms in the background.

---

[75] Not until the 1950's did the Alaska Road Commission have enough modern snow removal equipment to keep Thompson Pass open in the winter.

[76] Lt. Billy Mitchel of the U.S. Army Signal Corps surveyed many of the routes in the Territory. Years later, then Gen. Billy Mitchel would be considered by many as the true organizer of the U. S. Army Air Corps, the forerunner of the US Air Force. Olmstead, D. Lt. Col., 1933, Washington-Alaska Cable and Telegraph, *The Military Engineer*, Vol. XXV.

[77] John Kelsey grew up in Valdez and knew Whitham. Helping him load Nabesna gold and gold concentrate on steamships from the Valdez dock was another Valdez friend, Bill Egan. Years later Bill Egan was elected the first governor of the new State Of Alaska. Kelsey graduated from Stanford University, was a naval officer in World War II serving in the Battle of Leyte Gulf. Later he owned Valdez Dock Co., was mayor of Valdez, was chairman of the board of the Alaska Permanent Fund Corporation, Alaska's $26 billion savings account. He serves on boards of various companies including Alaska's largest locally owned bank in Alaska, First National Bank of Alaska.

[78] Al Ghezzy, born in Alaska, formed Alaska Freight Lines in the late thirties. His first truck was financed by Whitham to haul supplies to and gold concentrate out of the Nabesna Mine. His trucking company became one of the largest in Alaska.

[79] George Sullivan spent part of his youth at Kennecott when the mines were operating. At Valdez he worked on the dock with John Kelsey, loading Nabesna concentrate. He was a soldier in World War II and later was a U.S. Marshall, was manager of Alaska Freight Lines and was one of the most popular and the longest serving elected mayors of Anchorage. He always spoke highly of Whitham.

[80] Bob Reeve was a personal friend of Carl Whitham and named one of his sons Whitham. Bob Reeve served in World War I and later became a pilot and flew in South America. Later he became an Alaska bush pilot flying for many gold mines. During World War II he flew for the military. Following World War II he founded Reeve Aleutian Airways that operates today.

[81] Day, Beth, 1957, *Glacier Pilot*, The story of Alaska bush pilot Bob Reeve. Holt, Rinehart and Winston, NY.

[82] Father Hubbard S.J. studied many of Alaska's important glaciers and because of his research became known as the "Glacier Priest." Father Hubbard coined the term "Cradle of the Storms" for the Bering Sea. He knew persons around the world and interestingly was a friend of General George S. Patton, Jr. General Patton had received his fourth star when Father Hubbard paid him a personal visit at his Army Headquarters in Germany. Patton liked him and remarked in his crusty way "Talks very well when he forgets to advertise himself." Blumenson, M. (1974), *The Patton Papers*, Houghton Mifflin Co.

[83] Born in Pennsylvania, Charles E. Bunnell at age 22 came to the Territory in 1900. Was a teacher for the Office of Indian Affairs, principal of Valdez schools, member of Alaska Bar Association, appointed judge of the District Court in Fairbanks and on August 11, 1921 was elected by the Board of Trustees the first president of the Alaska Agricultural College and School of Mines, now the University of Alaska, Fairbanks. For 27 years he served as president. No one person was more responsible for development of the university than was he. To Judge Wickersham belongs the credit for establishing the Alaska Agricultural College and School of Mines, but it was Dr. Bunnell who built it into the University. He retired in July 1949. The regents conferred on him the title of President Emeritus. He continued to live on campus until his death in the fall of 1956. (For more see, Cashen, W.R. (1972), *Farthest North College President*, Univ. of Alaska Press.)

[84] Dr. John Patterson is a well-known and retired plastic surgeon in Fort Worth, Texas. During World War II he was a Naval officer and surgeon and served in the South Pacific.

[85] Nabesna Mine, because of its unique geology and accessibility, has long attracted university scholars and served as a geology field camp for a number of universities.

[86] Italian immigrants swelled the ranks of gold seekers and history tells Flex Pedroni, later Pedro, made the discovery that brought about the great Fairbanks gold field.

[87] The name of the Agricultural College and School of Mines was officially changed to the "University of Alaska" by the Territorial Legislature effective July 1, 1935. Therefore, the ski story must have taken place not later than the winter of 1935 before the name change.

[88] Dr. Skarland was a well-known skier, and at the University became an internationally known anthropologist. For years, Dr. Skarland was head of the Department of Anthropology.

[89] Harry Boyden was born in England and came to Alaska in the early days. He worked at the mine and after World War II, when the mine closed; he stayed on for a few years as watchman. "Boyden Creek" on Nabesna Road is named for him.

[90] Wayland, Russell G. "Gold Deposits Near Nabesna," *USGS Bulletin 933-B*, 1943. At the University of Minnesota Wayland studied under Prof. Grout. The University of Minnesota while Grout was Professor of petrography became world known for its research in the field of petrography and petrology. At Nabesna Wayland seems to have recognized, as Holdsworth earlier had, two periods of gold mineralization: early gold-pyrite-calcite and late gold-quartz.

CHAPTER 16

WORLD WAR II and the Final Days of Nabesna Gold Mine
By the spring of 1941, change was occurring in Alaska. Rumor of war was everywhere. Military construction was accelerating. The big military bases of Fort Richardson and Elmendorf Army Airfield at Anchorage were under construction. Ladd Army Airfield at Fairbanks (renamed Fort Wainright after World War II) was expanding. The 14-mile long Alaska Railroad spur and tunnel to the new port of Whittier drained the market of hardrock miners.[91] Those large government jobs swallowed up other laborers. Wages soared. For the first time since its startup, Nabesna Mine faced a shortage of miners such that Whitham was required to cut back scheduled work.

In early 1941, the Civil Aeronautics Authority (CAA), forerunner of the Federal Aviation Administration (FAA), set about to construct a network of Interior airfields in Alaska, primarily for military purposes. Those airfields were for Russian lend-lease warplanes. American Army pilots would ferry the war planes bound for Russia from Great Falls, Montana to Ladd Field at Fairbanks where they would be turned over to Russian pilots (some were reported turned over at Nome). The route from Montana to Alaska was across western Canada, and while airfields existed as far north as Whitehorse in Yukon Territory, none had been built on the 600 mile stretch between there and Fairbanks.[92]

To close that gap CAA selected as a "halfway" airfield a site along the Nabesna River in the Tanana Valley. The site was near the Nabesna Indian Village and the Native Village of Northway so, it is said, the new airfield was called "Northway." It was 60 miles down river from Nabesna Landing Field, which had years earlier been built by Nabesna Mine alongside the Nabesna River. The CAA's plan was to use Nabesna Landing Field as a staging point for its planned Northway field. The Alaska Highway in 1941 had not been built and no roads of any type reached the proposed Northway site. Nabesna Road was the closest of any road to Nabesna Landing Field. Nabesna Road and Nabesna Landing Field thus served as the vital link to get supplies and equipment to the proposed Northway site in the Tanana Valley. The CAA took over Nabesna Landing Field and used it as a staging field to airlift men and supplies from it to the new Northway site. To support that airlift, aviation gas, equipment and supplies had to get from Nabesna Road to Nabesna Landing Field. Bob Reeve in his book *Glacier Pilot* describes how that was done:[93]

> "The crated and barreled material was trucked from Valdez to Nabesna Mine, loaded on athey wagons–flat-track wagons with continuous tractor treads–and hauled by cat tractors over five miles of muskeg swamp to Bob's River Bar Landing Field." (meaning Nabesna Landing Field)

The tractors and athey wagons belonged to the mine, which leased them to the government. To begin with, Reeve flew the first two engineers to Tetlin Village on Tetlin Lake about 50 miles from the Northway site. He hired a riverboat, went down the Tetlin River to the Tanana River and then up stream to the Nabesna River and finally to the proposed Northway air field site. A crew of twenty Indians in six days hacked out of the brush an eight hundred-foot long rough airstrip. That was the beginning of Northway airfield. Morrison-Knudsen Co. (MK) was the contractor selected to build the new Northway airfield. It in turn subcontracted with Bob Reeve and others to do the airlift from Nabesna Landing Field to the proposed Northway field. According to Stan Cohen[94] Bob Reeve moved his family to Nabesna Landing Field for the summer of 1941. He and others flew equipment, men and supplies out of Nabesna Landing Field to the Northway site almost continuously from June to October. Most of that summer, like all summers, was nearly endless daylight.

The U.S. Government made arrangements with and used the mine as a staging area. Its truckers hauled equipment and supplies from Valdez down Nabesna Road where they off-loaded it at the mine. A fuel dump was established there. The mine provided truck crews and tractor operator room and board. As recently as 1999 one could find tangible evidence of this chapter at the mine. The National Park Service in 1999 removed some of the old barrels that fueled this operation. The first leg of the route used by the dozer operators pulling the loaded athey wagons from the mine to the Nabesna

One of the mine's smaller Athey wagons used by contractors to haul supplies and equipment from the mine to Nabesna Landing Field to support the Nabesna-Northway airlift by Bob Reeve and others.

Landing Field was along the old packhorse trail that ended at Orange Hill. At Jacksina Creek, the operators detoured north to the landing field. At the mine the dozer operators took a short cut to the packhorse trail across the mill tailings and dozed away a dike put there by the mine to retain the valuable tailings. In the many decades that followed, snowmelt and summer rains washed some of the tailing through the break and down slope.

"A horrible tractor trail," that's how Fred Bronnicke[95] described that early dozer-athey wagon route from the mine. Bronniche worked at the mine various times and later, after statehood, becoming road foreman at Slana Road camp.

Later, M-K hauled in its own bulldozers, scrappers and other heavy equipment, and off-loaded at about Mile Post 41 and took a new route to the landing field. M-K greatly improved and enlarged Whitham's "bush" strip and built a 6,000-foot long main runway and two 2,000-foot long cross-runways. According to Jim Rearden,[96] Nabesna Landing Field became one of the biggest in Alaska, shorter only than Elmendorf and Ladd Army Airfields. What had started as a landing field brushed out by Whitham for Nabesna Mine was now one of the few landing fields in the eastern interior that could handle bombers.

From June to October 1941, Bob Reeve and other pilots continued the airlift to the Northway site. According to Bob Reeve's daughter, Jance

Nabesna Landing Field originally built by the mine was enlarged many times to facilitate the airlift. To honor Bob Reeve the government renamed the field "Reeve Field" as it is known today. Following World War II Bob Reeve formed Reeve Aleutian Airways.

Reeve Ogle, they flew 1,100 tons of supplies and 300 workmen from Nabesna Landing Field to Northway field.[97] Jim Rearden[98] speculates the Nabesna-Northway airlift was "probably the largest airlift in the world to that time." Later in the summer MK organized a "Cat train" that pulled scrappers and other heavy equipment overland alongside Nabesna River from Nabesna Landing Field to the Northway site. The CAA christened Nabesna Landing Field "Reeve Field" and by that name it is known today.

The last military activity at Reeve Field seems to have been in the summer of 1945 a short time before the war with Japan ended. Ken Sheppard[99] a well-known Alaskan engineer, businessman and banker was then a Lt. Colonel assigned to Ladd Field, Fairbanks, as engineering officer. Relations with Russia by the summer of 1945 had cooled. Concern grew that if matters worsened the wartime-built, but no longer used, airfield at Port Clarence could serve Russian warplanes. Located near the western point of the Seward Peninsula, the airfield was closer to Russia than to Fairbanks. Colonel Sheppard was ordered to barricade the runway. The question then arose of other unmaintained but usable military airfields. According to John

Kenneth A. 'Ken' Sheppard, a prominent Alaska engineer and businessman served as Lt. Colonel durring World War II. In 1945, his was the last military report on Reeve Field.

Hellenthal,[100] a major during the war, Colonel Sheppard was ordered to review the situation at Reeve Field. Sheppard reported that if the Russians got to the airfield without being shot to pieces the watchmen at the mine could handle the situation.

In 1941, the mine closed early in the fall and Whitham took the steamer to Seattle. On December 7, 1941, Japan bombed Pearl Harbor and Congress declared war against Japan. World War II began for the United States.

The war effort would hit the labor market hard in Alaska. Knowing miners were hard to come by the year before (1941) the United States declared war; Whitham guessed the coming season (1942) would be even bleaker. When word spread in Seattle that the Army would restrict civilian travel to Alaska except to work for military or defense projects, Whitham knew shipping restrictions on fuel and other supplies would follow. In fact by March 1942, the Alaska Steamship Company removed some of its larger vessels from the Alaska Civilian Service for military use. Word from Fairbanks was that Fairbanks Exploration Company, called "FE" and the largest gold dredging company in Alaska, was likely to shut down its dredges for the duration of the war.[101] The handwriting was on the wall. On March 20, 1942, Whitham sent a telegram to Claude Steward, vice president of Nabesna Mining Corp. in Chitina, Alaska, bearing the following message:

> "Conditions make advisable we close Nabesna Mine
> for duration of war. Please notify any former employees
> up there to this effect".

The government officially closed all gold mines on October 8, 1942, when the War Production Board (WPB) issued Order L-208. That Order prohibited new gold developments after October 15, 1942 and ordered all gold mines to cease active mining within 60 days, allowing for only maintenance.[102] Whitham had no choice but to wait out the war in Seattle.

> **TELEGRAM — Signal Corps, United States Army, Alaska Communication System**
>
> Mr. Claude Stuart
> Valdez Alaska
> Forward to Chitina Alaska
>
> Seattle Wn
> March-20 1942
>
> Conditions make advisable WL close Nabesna Mine for duration of war please Notify any former employees up there to this effect
>
> Carl F. Whitham
> c/o Harold L. Scott & Co
> Insurance Bldg,
> Seattle Wn
> phone Ma. 4810
>
> pd Total
> $2.69

Whitham was in Seattle when Pearl Harbor was bombed and World War II broke out. he closed Nabesna Mine for the duration of the war and sent the above telegram to Claude Stuart, then a company director and storekeeper at Chitina.

By the end of World War II, Whitham had come around to accepting Holdsworth and Joralemon's recommendations. On returning to the mine he set about to explore and develop the Bear Vein and Tower Knob Fault and the area on the west side of Nabesna stock. In that western area he intended to explore for the source of gold found in Discovery and Stamp Mill Gulch.

To fund the anticipated exploration, the company in 1946 issued new stock through the Seattle Exchange. It was oversubscribed. In the summer of 1946 there was reason to believe the mine could become even larger than in pre-war times. Whitham notified the Tacoma Smelter that the mine planned to start regular shipments in 1947 and asked for its smelter schedule and fees. His friend Bob Reeve had moved from Valdez to Anchorage, and had organized Reeve Aleutian Airways. Being busy with military contracts out in the Aleutians, he was no longer flying into the Nabesna country. Therefore, he contacted his friend Merle "Mudhole" Smith of Cordova

September 14, 1946

Tacoma Smelter
American Smelting & Refining Co.
Tacoma, Washington

Gentlemen:

    We are hauling our first load of concentrates from post war mining to Valdez next week and we will be making shipments of concentrates to you for smelting as fast as we can accumulate shipments. We will appreciate your giving us a contract for smelting at a figure as near as possible to the rate charged in our last contract with you.

    Bennett's Chemical Laboratory will represent us the same as before and we will write both you and Bennett's each time shipment is made from Valdez.

    The entire proceedure will be exactly the same as when we shipped you in the past. Net proceeds of each shipment will be handled through Pacific National Bank, Seattle, Wash. for credit of First Bank of Cordova, Cordova, Alaska for account of Nabesna Mining Corporation, Nabesna, Alaska.

    Yours very truly

    NABESNA MINING CORPORATION

By_____
Carl F. Whitham, President

CC:
  Bennett's Chemical Laboratory, Tacoma, Wasington

  Pacific National Bank, Seattle, Washington

  First Bank of Cordova, Cordova, Alaska

  Gentlemen:
    We will appreciate the same efficient handling of our account as in the past

    Carl F. Whitham

After World War II, Whitham opened the mill for several months in the summer of 1946. In September he sent a small shipment of concentrate to Tacoma Smelter. He notified Tacoma Smelter he would resume operation in 1947. That was not to be and the summer of 1946 would be his last in Alaska. Carl F. Whitham died in Seattle in February 1947.

Airlines on behalf of Whitham to establish a client relationship between the mine and another carrier. Cordova Airlines was establishing scheduled service into Fairbanks and other Interior stops.

Post-war defense work was at a high pitch in Alaska, and Russia was then the threat. Whitham could not find enough miners in the summer of 1946 to start his planned exploration work. He was able to hire only a few men needed to operate the mill one shift per day. They milled a few tons of stockpiled ore and some low grade from a nearby prospect. Even though Whitham and others knew that was not profitable it got the mill opened and operating.

In letters and articles by Whitham, he made known his awareness that mining must change if it were to operate and compete in the post war cost plus government defense spending in Alaska. With new objectives and plans in place, he believed the mine could operate profitably. He shut down the mill in September and made it ready to re-open in the spring of 1947.

When Whitham left for Seattle in October of 1946, he could not have known that would be his last summer in Alaska. Four months later in early February 1947, Carl F. Whitham died in Seattle. One would guess that Whitham, if he had the capacity to have feelings about his death, his prime regret was that he would not be able to honor the trust his old and new shareholders had reposed in him.

Whitham had found, built and run Nabesna and like many "one man managed" mines there was no real second in command. None of the secondary company officers were mining men and no one was left behind to take over. One can only speculate about Nabesna Mine had Carl F. Whitham lived. However, men and times had changed. The mine, with all the potential Carl Whitham, Phil Holdsworth and others saw in it, never reopened.

---

[91] Swalling, Al, 1999, *Oh, To Be Twenty Again–and Twins*, A&M Publishing, Anchorage, Alaska.
[92] The airlift seems not to have one official name and is referred to as The Northwest Ferry Route, Alaska-Siberia Lend Lease to Russia, Alaska-Siberia Air Route and The Northwest Staging Route.
[93] *Glacier Pilot*, op. cit.
[94] Cohen, Stan, 1992, *The Forgotten War*, Pictorial Histories Publishing Co, Missoula, Mont.
[95] Fred Bronniche and his wife Inze came to Alaska in the mid-thirties and mined on their own for a few years. Fred was a chemist and worked off and on at the mine as an assayer and engineer. Later he worked for ARC during and after World War II. After Statehood he continued as State Road foreman at Slana.
[96] Rearden, Jim., *In the Shadow of Eagles*, Rudy Billberg's story, Alaska Northwest Books, 1992.

[97] Ogle, Jance Reeve, 1993, *Reeve Aleutian Airways*, in Alaska at War Committee, 1995.
[98] *In the Shadow of Eagles*, Op.cit.
[99] Ken Sheppard, a 1922 civil engineer graduate of Leheigh University had many years of civil and military design and management experience. Following Pearl Harbor and well beyond draft age he enlisted in the Army and was sworn in as a captain. As he would later say, "No reason to go in as a lieutenant." At war's end he stayed in Alaska and became active in many developments throughout the Territory and later the State. In later years he was active in banking. Sheppard or William Tobin, then editor of the *Anchorage Times,* coined the popular saying "termination dust" when snow on the mountains signified the end of the summer construction. Ken Sheppard died in 1995.
[100] John Hellenthal was raised in Juneau, Alaska where his father served as a judge. He met many prominent persons visiting his parent's home. One was Professor Bateman of Yale and at the time consulting geologist for Kennecott Mines. Carl Whitham was another visitor. Hellenthal retained an interest in geology and mining. He graduated from Notre Dame University and after the war became a well-known lawyer in Alaska and the Pacific Northwest. Among other positions he served on the Alaska Constitutional Convention that paved the way for Statehood. His clients did not always understand his eloquence in court. A case in point was when representing Strandberg Mines, the second-largest gold placer producer in Alaska, its president Harold Strandberg, a hard-boiled mining engineer and life long Alaskan (and a life long friend of Hellenthal), complained to a friend "Damnit! Who's he arguing for?" Hellenthal won the case. John Hellenthal died in 1989.
[101] Spence, C.C. *Fairbanks Exploration Goes to War*, Alaska Historical Society, Vol. 10, 1995.
[102] The War Production Board believed closing the gold mines would release as many as 3000 miners to work in the iron and copper mines. Some reports claim no more than 100 did, with the bulk going to work in shipyards etc. On July 1, 1945, Order L-208 was lifted. United States was the only country in World War II that closed down it's gold mine industry and Congressman Clair Engle of California said the order wrecked a $200 million industry. The gold industry sought compensation but bills went nowhere in Congress. In 1955 a number of western mining companies took their claim to the Court of Claims in Washington. The argument questioned WPR power to close one industry so as to shift labor to another. The mining companies called it excess in eminent domain and the closed gold mines were entitled to compensation. The Court of Claims agreed that "just compensation" for profits lost during the close down be paid. In 1958, the United States Supreme Court reversed the Court of Claims decision, saying gold mining operations had only been "suspended" and the property had not been confiscated.

CHAPTER 17

The Early Cold War Period

Articles in several newspapers[103] and the *Army Times*[104] report Nabesna Mine likely played a role for the military as an underground arms cache during the early "Cold War" period. The articles tell how military officers during the late 1940's and early 1950's knew the reduced number of post-war armed forces in Alaska could not repel a serious Russian attack on Alaska. Mostly World War II veterans, the commanders knew underground resistance had played a role in every theater during World War II and they were not blind to the role resistance might have to play in Alaska. The articles suggest the U.S. Army Counter Intelligence Corps (CIC) set about to organize an underground force that would remain behind in Alaska if the Russians invaded and the small American army had to withdraw into Canada.

This was only five to six years after the greatest war in history. The underground resistance that had such a heroic part of the war was fresh in everyone's mind. No doubt the assumption of the military was that underground cells could be effective in any new conflict with new enemies. CIC knew if underground bases were to be effective they must be strategically located along transportation routes but not obvious to locals and least of all to Russian invaders. Nabesna Mine fit that scheme of things. The articles suggest other bases were located elsewhere in Alaska. The articles go on to say the underground members of the cadre selected were of high standards and knew the local area and people. Members of the so-called "Nabesna faction" were from the Nabesna-Slana area and were secretly sworn-in by CIC officers at Fort Richardson Army base at Anchorage. The military covertly transported to and cached at the mine arms and other items to support an underground cell.

As the cold war lingered on and the threat of invasion appeared less likely, the "Nabesna faction" only then learned the cold fact that they had been setup as a decoy. CIC, the articles speculate, knew the Russians, being great underground fighters themselves, would expect cells in Alaska and would not stop until they rooted them out. Word would be leaked to the Russians so as to allow them to find and neutralize the Nabesna faction. Having done so, CIC believed the Russians would relax their search and a more secret underground could then operate with a greater degree of safety.

A look back to the late forties and beginning of the Cold War is worthwhile in assessing the truth to this story. The Soviet Union following World War II dominated Eastern Europe and much of the Far East. By and large the American public, tired of the war, viewed Russia as a wartime friend and remained unaware, or unbelieving, of reports of the Soviet Union's growing belligerence towards America and its naked thrust to dominate Europe. That feel-good image was thrown aside by General Mark Clark, then military governor of Austria, when asked to speak on NBC. He ripped into the Russians and told of their "rape and pillage of Austria, double dealing with their allies and intransigence at postwar conferences."[105] General Clark was not the only one worried by Russia's actions. In 1947, the Joint Chiefs of Staff created the Alaska Command, ALCOM, and the first-ever joint command that placed all military forces in Alaska and the North Pacific under the first unified command.[106]

By 1949 Russia had developed the A-bomb and had done so four years before Washington believed it could. In addition it had the air power to deliver the bomb. That delivery route would be over Alaska. During World War II Alaska was the military bastion of the North, but by 1949 the troop strength was reduced to less than 20,000 men. As early as 1948 Herb H. Hilscher in his book *Alaska Now* wrote, "If a polar war started right now, a high-ranking Army officer told me, we would not be able to defend Alaska."[107] By that time mainland China had fallen to the communists and Russian bombers were probing Alaska air space. In November 1949 *U.S. News and World Report* published an article entitled, "Alaska–Another Pearl Harbor?" It warned that Alaska's military unpreparedness had left the door open for the Russians. Then in the spring of 1950 North Korea, supported by Russia, invaded South Korea.

Those combined events caused Alaska Territorial Governor Ernest Gruening to press Washington for more military presence in Alaska warning "A Russian paratroop division could take Alaska in a week."[108] The Russians would target Fairbanks and Anchorage, Governor Gruening warned, and with Russia operating from those military airfields, all of the U.S. would for the first time come within range of Russian bombers. He tells how Secretary of the Air Force Stuart Symington was concerned, too, and was prepared to move the Boeing plant out of Seattle and across the Rockies where it would be a more difficult target for Russian bombers. Needless to say, the Seattle Chamber of Commerce vehemently objected and the plan was scrapped. Those were unsettling times, and to believe the military commanders in Alaska did not take a Russian threat seriously or make contingency plans is to view them as amateurs. Such they were not.

Another factor lending credence to the reports is that few other remote mines in Alaska, by happenstance or otherwise, had so close a relationship with so many important Army officers, as did Nabesna. That relationship went back to Steese, Elliot and Edgerton who were generals by World War II. During the Northway airlift a new crop of Army officers at Fort Richardson and Army Air Corps officers at Elmendorf and Ladd Field came to know Nabesna. Bob Reeve knew Nabesna, and in post-war Alaska he had close friendships with the top military commanders. What better authority was there if one wanted to know where to cache military related things? As someone once said "the Army likes to go to places it knows." And the Army knew Nabesna.

Long afterwards, in the seventies, persons doing mapping at the mine stumbled onto one of the old turn-of-the-century, Royal Development drifts. The drift had no mineral importance, and there was no reason for any one to look for it. It is located low on the mountain where the slope is grown up in thick brush and trees. The drift is only 30 feet deep and never used by Nabesna Mining Corporation. Still, scattered in the brush near the portal these persons found broken crates and boxes, and not the type seen elsewhere at the mine. No less interesting was that fixed to the portal was a door and from it hung a broken clasps and padlock. The mine did have heavy wooden doors fixed to portals of working drifts. This was to stabilize temperature and keep wild things out, but none were ever padlocked. However, remember this old drift was never used by the mine. There was no reason for a door to be there in any case. There was even less reason for a padlocked door. The door seems built of scrap boards and doesn't have the look of any other portal door at the mine.

Someone, so it seems could have placed things in there, put a door at the portal and then locked it. No less puzzling is a small white and blue sign pasted to a corner of the door that read "Government Property Keep Out." In the summer of 2001 the sign was still on the door but the words had faded.* One can only speculate that maybe years ago someone without a key because it had been lost, broke in, cleaned the place out and left the wrappers behind. The articles speculate more caches may be hidden there?

Photo taken by Whitham in 1923 of an old drift located low on the mountain, probably driven by early prospectors before 1910. The portal has no door. Nabesna Mining Company never used this drift.

Photo of same drift taken by persons who "rediscovered" it in the 1980's and shows a door fixed to the portal. On the door is a broken latch and padlock. A small white and blue sign on the door reads "Government Property Keep Out." Rediscovered empty crates were found scattered about but the drift was empty. Several newspaper articles in the 1970's described that at the beginning of the Cold War in the late 1940's and early 1950's U.S. Army Counterintelligence used Nabesna Mine, as well as other places, as a resistance base and arms cache. This old drift could have been one cache.

* During the mid 1990's the National Park Service learned of the "arms cache" story, and fearing live ammo and hand grenades might still be there, notified the Army Corps of Engineers. An Army engineer spoke with the owners of the mine and seemed not impressed with the story until mention was made of the small "Keep Out" sign. "Ah!" That kindled interest in the skeptical Army Engineer. "That's what a "spook" or government bureaucrat would do. Cache the arms, put a door on, lock it and then slap a sign on warning, "Government property keep out."

[103] "Nabesna Road: Precarious Trail to Nowhere," *Anchorage Times,* Anchorage, Alaska, 1979. And "Nabesna Faction," *Anchorage Great Lander,* Anchorage, Alaska, 1980.

[104] *Army Times,* April 1982. A short article tells how owners of Nabesna Mine gave the 172nd Infantry Brigade permission to use the mine for training. Soldiers from the Northern Warfare Training Center at Fort Greely used the cliffs for advanced climbing under command of Colonel Thomas P. Leavitt, U.S. Army and Colonel John Kaiser, U.S. Marine Corps. The article goes on to say the U.S. Counterintelligence formerly used the mine property as an underground base for an arms and supply cache.

[105] Mosley, Leonard, 1982, *Marshal, Hero for Our Times*: Hearst Books, NY.

[106] Strobridge, Truman R., (1966) *STRENGTH IN THE NORTH: The Alaska Command 1947-1967,* Elmendorf Air Force Base, Alaska.

[107] Hilscher, Herbert H., 1948, *Alaska Now*: Little Brown and Company, Boston.

[108] Gruening, E., 1973. *Many Battles*, The Autobiography of Ernest Gruening, Liveright, New York. Dr. Gruening was by training a medical doctor. President Franklin D. Roosevelt apointed him Governor of the Territory of Alaska in 1938. he served as governor of the Territory until 1954. On Alaska Statehood he was elected United States Senator of Alaska in 1960.

CHAPTER 18

Conclusion

The Nabesna, at the turn of the 20th century, when first seen by Europeans, was a place "back of beyond." Today's traveler along Nabesna Road would still agree. He or she views a landscape little changed since first seen by the Batzulnetas Indians as they ventured there after the ice age. It has changed hardly at all since Carl F. Whitham first saw it so many years ago. About all Carl Whitham would say is that the trees have grown. Those great white cliffs towering above the old mine town have served as a landmark over the ages and they hold secrets which will not likely ever be revealed about the land itself, as well as the men who, in the last second of geologic time, appeared there. Nabesna is truly a place of majesty and mystery.

BIBLIOGRAPHY

Allen, Henry T., LT., 1887, *Report of An Expedition to the Copper, Tanana and Koyukuk Rivers, of the Territory of Alaska in the year 1885*: Washington; Government Printing Office.

Atwood, Evangeline, 1979, *Frontier Politics, Alaska's James Wickersham*: Binford and Mort, Portland, Oregon 97202

Author unknown, 1982, *Army Times*.

Author unknown, 1980, "Nabesna Fraction": *Anchorage Great Lander*, Anchorage, Alaska.

Ballou, Maturin M., 1898, *New Eldorado*: Houghton Mifflin and Co., Boston.

Blumenson, M., 1974, *The Patton Papers*: Houghton Miffin Co., Boston.

Brooks, Alfred H., 1953, *Blazing Alaska Trails*: Univ. of Alaska.

Cashen, W.R., 1972, *Farthest North College President*: Univ. Alaska Press.

Cohen, Stan, 1992, *The Forgotten War*: Pictorial Hortorise Pub. Co., Missoula, Montana.

Day, Beth, 1957, *Glacier Pilot, The story of Alaska bush pilot Bob Reeve*: Holt, Rinehart and Winston, NY.

Douglass, Wm. C., 1964, *History of Kennecott Mines*: Seattle, Washington

Elies, Scott A., 1995, *Ice Age History of Alaska National Parks*: Smithsonian Institution Press, Washington, D.C.

Eppinger, R.G., et al., 1995, *Geochemical Data for Environmental Studies at Nabesna and Kennecott, Alaska*: USGS, 1st Draft Version, Open File, Denver, Colo.

Fagin, Brian M., 1987, *The Great Journey, the peopling of ancient America*: Thames & Hudson, NY.

Gillete, Helen, 1979, "Nabesna Road A Precarious Trail to Nowhere": *Anchorage Times*. Anchorage, Alaska.

Gruening, E., 1973, *Many Battles, The Autobiography of Ernest Gruening*: Liveright, New York.

Hilscher, Herbert H., 1948, *Alaska Now*: Little Brown and Co., Boston.

Hopkins, David M., 1973, "Sea Level History in Beringia during the last 250,000 years": *Quaternary Research*, Vol. 3.

Janson, Lone E., 1975, *The Copper Spike*: Northwest Publishing Co., Anchorage, Alaska.

Kirchhoff, M.J., 1993, *Historic McCarthy*: Juneau, Alaska.

Meyer, M.P., 2000, *Mineral assessment of Ahtna, Inc. selections in Wrangell-St. Elias National Park and Preserve, Alaska*: Final Report No. 34, Dept. of the Interior.

Moffit, F.H., 1910, *Mineral Resources of the Nabesna-White River district, Alaska*: USGS Bull. 417. Washington, D.C.

Moffit, F.H., 1943, *Geology of the Nutzotin Mountains, Alaska and Gold Deposits near Nabesna*: USGS Bull. 933-B.

Mosley, Leonard, 1982, *Marshall Hero For Our Times*: Hearst Books, New York.

Ogle, Jance Reeve, 1993, *Reeve Aleutian Airways in Alaska at War 1941-1945*: Anchorage, Alaska.

Olmstead, Dawson, Lt. Col., 1933, "Washington-Alaska Cable and Telegraph": *The Military Engineer*, Vol. XXV.

Pierce, R.A. and Alton S. Donnelly, 1978, *A History of the Russian-America Company*: Univ. of Washington Press, Seattle.

Pilgrim, Earl R., 1975, "The Treadwell Mines": *The Alaska Journal Quarterly*, Vol. 7, No.4.

Pilgrim, Earl R., 1931, *Nabesna Mining Corporation, Whitham group: Report on cooperation between the Territory of Alaska and United States in making mining investigations and in inspection of mines for the biennium ending March 31, 1931*: Juneau, Alaska.

Rearden, Jim, 1992, *In the Shadow of Eagles*: Alaska Northwest Books, Anchorage, Alaska.

Richter, et al., 1990, *Age and Progression of Volcanism, Wrangell Volcanic Field, Alaska*: Bulletin of Volcanology, Spring 1990.

Richter, et al., 1995, *Guide to the Volcanoes of the Western Wrangell Mountains, Alaska-Wrangell-St. Elias National Park and Preserve* USGS Bull. 2072.

Ricks, Melvin B., 1965, *Directory of Alaska post offices and postmasters, 1867-1963*: Ketchikan, Alaska, Tongass Pub. Co.

Seabrock, John, 1989, "Reporter at Large": *New Yorker*, New York.

Spence, C.C., 1995, "Fairbanks Exploration Goes to War": *Alaska Historical Society*, Vol. 10.

Strobridge, Truman R., (1966) *Strength in the North, The Alaska Command 1947-1967,* Elmendorf Air Force Base, Alaska.

Swalling, Al, 1999, *Oh, To Be Twenty Again–and Twins*: A and M Publishing, Anchorage, Alaska.

Tower, Elizabeth A., 1988, *Big Mike Heney, Irish Prince of the Iron Trails Story of Copper River Northwestern Railroad*: Anchorage, Alaska.

Tower, Elizabeth A., 1990, *Ghost of Kennecott, the story of Stephen Birch*: Anchorage, Alaska.

Tower, Elizabeth A., 1991, *MINING MEDIA MOVIES: Cap Lathrop's Keys for Alaska's Riches*: Anchorage, Alaska.

Tower, Elizabeth A., 1996, *Ice Bound Empire*: Anchorage, Alaska.

Wickersham, James, 1938, *Old Yukon: Tales-Trails-Trials*: Washington Law Book Co., Washington, D. C.

GLOSSARY

Adit: Generally a short underground working that may start from the surface or from a drift or tunnel.

Athey wagon: A flat bed wagon with tractor tracks instead of wheels to haul heavy loads across marshy ground or snow and pulled by a bulldozer.

Ball Mill: A large 50 ton closed tube-shaped equipment that rotates slowly, allowing 3 to 4 tons of iron balls to grind ore to the size of table salt or finer.

Concentrating Table: A five-foot wide by 16-foot long table that jerks back and forth and fixed with riffles to collect raw gold and discard waste.

Concentrate: Any crushed ore from the milling process that contains value in gold or other metal.

Cross-cut: Generally a horizontal passageway between two drifts or other underground workings.

Diamond Drill: A drill using hollow drill pipe to which a hollow bit is screwed to the end. The edge of the bit is impregnated with industrial diamonds. Rotating at high speed the diamond bit will drill through hard rock. As it does, a core of that rock passes up through the hollow bit and into a core barrel. The core barrel is retrieved and the miner can examine the core and tell the type of rock or mineralization drilled.

Drift: Usually any underground working opened to the surface. When opened to the surface at both ends it's called a tunnel.

Carbide lamp: A miner's lamp fixed with a small bowel filled with dry calcium carbide that when drops of water are added acetylene gas is produced. When ignited the gas burns with a bright light.

Flotation: Method used at Nabesna and elsewhere to separate gold from other minerals. When common pine oil is added, as at Nabesna, to a tank filled with water and agitated by a propeller, a froth of pine oil forms. Gold but not pyrite sticks to the froth. The pyrite sinks to the bottom and is discarded into the tails. The froth floats to the surface where it is scraped off and the gold collected.

Level: This is the general name for a horizontal underground working. Level and drift are often used interchangeably.

Lode: A mineral bearing body of valuable ore that unlike a well-defined vein has an irregular shape such as a pod.

Magnetite: This mineral is a black, hard iron mineral that will attract a compass needle. Magnetite is common at Nabesna and always contains some gold values.

Outside: An expression once common but less used today meaning any place in the lower states, such as "They went outside to Seattle."

Portal: The surface entrance to an underground working.

Round: When miners are advancing an underground working such as a drift a series of holes are drilled in the face, loaded with dynamite, and then blasted. When all the holes are completed it is called a round. As in "They shot that round" meaning the miners set off the dynamite.

Shaft: A shaft is a vertical or steep incline opened from the surface. The only shaft at Nabesna is the 50 foot deep shaft Whitham and Kansky sank on the Bear Vein outcrop.

Skarn: Skarn is a geological term that includes a suite of minerals formed under high temperature conditions called, "contact metamorphic." At Nabesna, garnets and magnetite are the most easily recognized skarn minerals.

Strike & dip: The compass bearing of a vein is the strike such as "The strike of the vein is S 50 W." The inclination of the vein from the horizontal is called the dip such as "The vein dips 25 degrees west into the mountain."

Stope: A man-made working where ore is mined above and accessed from a drift or level is called a stope. Almost all ore at Nabesna was mined by stoping.

Tailings: The purpose of milling is to separate gold or other metal from unwanted minerals. The unwanted minerals are called tails. Milling is generally 95% efficient but sometimes much less. The early mill efficiency at Nabesna was only 50% (50% of the gold was lost to the tails).

Tunnel: An underground working that goes in one side and comes out the other side of a mountain or hill is a tunnel. Often the word tunnel is used for any underground working that is entered from the surface. At Nabesna the "Nugget Tunnel" is technically a drift.

INDEX

A-bomb, (Cold War), 130
Abercrombie, Capt. Wm. R. (Early Army explorer in Alaska), 24, 27, 34, 89
Alaska Agricultural College and School of Mines (the University of Alaska Fairbanks), 62-63, 80, 90, 113, 118
Alaska Communication Services (part of the U.S. Army Signal Corps), 103
Alaska Highway, 2, 46, 89, 119
Alaska Nabesna Corp. (early owners of Orange Hill claims), 58
Alaska Now (book by the late Herb H. Hilscher), 130, 133, 138
Alaska Peninsula, 19
Alaska Road Commission (ARC), 43-44, 49, 54, 57-58, 100, 117
Alaska Railroad, 119
Alaska Steamship Company, 39, 44, 71, 123
Alaska Syndicate (Original owner of Kennecott Copper Mines), 65
Alaska Weekly (Newspaper published in Seattle), 45, 66, 68
Aleutian Chain, 19
Algal mat (an organic mat found between dolomite and limestone at Nabesna), 9-10, 12-14, 19
Allen, Lt. Henry T. (First European in 1885 to see Batzulneta & White Mountain–at what is now Nabesna), 6-7, 24-28
American Exchange-Irving Co (sued James J.Godfrey), 66, 68
Apache (words and numbers used by Batzulneta Indians much like Apache), 6, 26, 28, 49
Arctic Ocean (froze 2-3 million years ago), 19
Army Times (newspaper published by U.S.Army), 129, 133, 137
Ahtell Creek (located in Slana area), 31-32, 34-35, ,40, 45, 66
Athey wagon (wagon with tracks used to haul supplies from mine to Reeve Field), 120-121, 141
Auriferous (gold combined with pyrite), 14, 17

Bateman, Prof. Alan M. (internationally known geologist and consultant for Kennecott), 42, 45, 50, 127

Batzulnetas Village and people (Village of warlike Indians who likely saved Lt. Allen's. expedition), 5-7, 17, 20-22, 24-28, 32, 34-35, 40-42, 49, 56, 135

Batzulneta, Chief (Chief of Batzulneta people at time of Lt. Allen), 24, 26

Baldwin, Asa C. (well know mining man & mineral surveyor), 83, 90

Bering land bridge, (formed during last Ice Age), 20

Bear Vein (rich gold vein that made possible Nabesna mine), 2, 12, 22, 41-54, 58, 61-63, 72-78, 80-81, 83-85, 88-89, 96, 124, 142

Birch, Stephen (first president of Kennecott Copper Co.), 44-45, 50, 140

Board of Missions Protestant Episcopal Church (helped establish schools), 47, 49

Boeing (aircraft co. in Seattle), 131

Boyden, Harry, (early day resident of Nabesna), 115, 118

Bonanza Mine (Kennecott Copper Mine), 33, 65

Brooks, A.H. (geologist and later director of USGS), 27-28, 49, 137

Bronniche, Fred and Inze (Long time residents of Slana), 115, 121, 126

Bunnell, Charles E. (First president of Univ. of Alaska Fairbanks), 106, 108, 118

Bureau of Land Management (BLM), 104

Bureau of Mines, 97

Cabin Creek (water source for Nabesna), 14-15, 40, 88, 102

Caribou-Comstock claims, 40-42

Chitina, 32-33, 37, 39, 40-41, 58-59, 61, 71, 84, 110, 112, 123-124

Chistochina, 20, 49

Chistochina River, 30, 39

Chilkoot Pass, 29

Clark, Gen. Mark

Civil Aeronautics Authority (CAA instrumental in Nabesna-Northway airlift), 1, 119

Cohen, Stan (author of history books concerning Alaska), 120, 126

Cold War, 2, 129, 130, 132

Colliers.(a one time popular magazine but no longer published), 45

Columbia School of Mines, 65

Copper River, 20, 23-27, 32, 37, 39, 40, 47-50,71, 84, 102, 107-107, 140

Cordova, 37, 39, 41, 43, 61, 68-71, 84, 103, 106-107, 111, 114, 124, 126

Cordova Air Service, 103

Cordova Daily Times, 70-71

Copper River & Northwestern Railroad (served Cordova, Chitina and Kennecott Mines), 106

Counter Intelligence Corps (CIC could have cached arms at Nabesna during Cold War), 129
Curb and Mining Exchange (Seattle), 63

Day, Beth (author of *Glacier Pilot*), 118, 137
Dall sheep, 20, 22
Dawson, 29
Devils Mountain Lodge (well known lodge on Nabesna Road near mine), 17, 117
DeWitt, Angus (present owner of historic Slana Roadhouse on Nabesna Road), 20, 49
DeWitt, Lawrence (he built Slana Roadhouse and delivered mail by dog sled), 47, 61, 103
Discovery Gulch (where placer gold first discovered at Nabesna 1898), 88
Do Drop Inn (Saloon and business house at Nabesna), 110-113
Douglass, Wm. C. (engineer for and wrote history of Kennecott Mines), 69

Eagle City on Yukon River (end of Eagle Trail), 34, 60, 103
Eagle Trail (trail staked 1903 between Valdez and Eagle on Yukon), 31-33, 35, 39, 44, 57, 103
Edgerton Highway (between Chitina and Richardson Hwy.), 29, 39, 84, 135
Egan, Bill, (First governor State of Alaska), 117
Eldorado Creek (gold placer worked by Whitham at Chisana), 27, 36
Elliott, Major, 54, 60
Elmendorf Army Airfield, 1, 19, 121-, 131, 133, 139
El Se Ba (reported as old Indian name for cliffs at Nabesna), 5, 7, 21, 28
Environmental Protection Agency, (EPA), 97

Fairbanks Exploration Co. (called "FE" & once Alaska's largest placer gold producer), 123, 127, 139
Federal Aviation Administration (FAA), 119
First Bank of Cordova, 68
Flat, 44
Fort Richardson, 119, 129, 131
Fort Selkirk, 23

Gakona Roadhouse (historic roadhouse on Glenn Hwy.), 39, 40-41, 71, 80, 84
Gamblin, Sam (Rambling Sam), 87-88
Garnets, (semi precious minerals common at Nabesna), 13, 17, 22, 142

General Land Office (GLO), 32, 104
Gillam, Harold (well known Alaska bush pilot who flew gold from Nabesna), 78-79, 95, 103, 114
Glacier Pilot (book about Bob Reeve), 105, 118, 120, 126, 137
Glenn Highway (also called Slana-Tok Cutoff or just Cutoff), 2, 5, 15, 29, 34, 39, 41, 46, 89
Glover, A.L.(Seattle assayer), 41, 63, 98
Goat trail (dangerous section of trail between McCarthy and Chisana), 34, 45
Gold, 1-3, 5, 7-8, 12-14, 16-17, 22, 25, 27-32, 34-39, 41-45, 50-54, 57, 60-61, 63-66, 68-69, 71, 77-78, 82, 90-91, 93-98, 100-101, 104, 111, 114, 118-119, 123-124, 127, 141-143
Great Falls, Montana (start of the route lend-lease warplanes took to Fairbanks), 119
Godfrey, James J. (sued Whitham claiming interest in Nabesna Mine), 33, 39, 65-69
Gruening, Governor Ernest (territorial governor and Alaska's senator upon statehood), 106, 130, 133, 138
Guggenheim (involved with financing Kennecott mines), 65
Hellenthal, John (long time Alaska lawyer), 123, 127
Herbert, Charles F. (well known mining engineer and former Deputy Commissioner of Natural Resources), 90
Hilscher, Herb H. (author of *Alaska Now*), 130, 133, 138
Holdsworth, Phil R. (outstanding mining engineer at Nabesna, Philippines and first Commissioner Alaska Department Natural Resources), 81, 83, 85-86, 90, 95-98, 116, 118, 124, 126
Honorably discharged (Whitham World War I), 39, 66
Hot Springs, Ark., (Marie Thornton and Whitham were married 1919), 39, 66
Hubbard S.J., Father Bernard (well known Jesuit priest and geologist who stayed at Nabesna and studied Nabesna Glacier), 105-106, 118
Hubbell, A.H. (Editor of *Engineering & Mining Journal*), 53
Hudson Bay Company, 23, 28

Ice Age, 19-21, 135
Idtarod, 44, 49

Jack Creek , 15, 47
Jack Lake (bush pilot Gilliam used lake in winter to land his ski plane on), 71, 105
Jacksina Creek, 5, 8, 14-15, 25-26, 105, 121
Joralemon, Ira B. (consulting geologist), 89-90, 116, 124
Jumbo Mine (Kennecott Mine), 33, 65, 69

Juneau, 39, 41, 44, 127, 138-139
Justin, Willson (born near Nabesna and former board member and President of Ahtna Regional Native Corp.), 20, 49

Kamchatka Peninsula, 19
Kay Copper Corporation (managed by Godfrey in Ariz.), 66-67
Kansky, Steve (at Bear Vein used rocker to recover free gold), 53-54, 56-57, 142
Keenan Peak (Mountain near Nabesna Mine), 5
Kelsey, John (long time Alaskan businessman and during college years hauled gold bars and concentrate from Nabesna mine to Valdez), 104, 117
Kennecott Copper Corp., 33-34, 44, 65, 84, 101
Kennicott (present town), 34, 37, 45, 67, 90, 106, 108, 117, 138
Kenny Lake, 61
Ketchikan, 34, 139
Kirchhoff, M. J. (author of *Historic McCarthy*), 35, 41, 138
Klondike, 28-30
Knight, Earl (Publisher of *Alaska Weekly*), 68
Kuhltan (said to be early Indian name for Mt. Sanford), 28

Lathrop, "Cap."Austin, 86-69, 140
Liberty (popular magazine but went out of business after World War II), 45
Lyle Airways, 103
Ladd Army Airfield (now Fort Wainright), 119, 121-122, 131
Land bridge, 20

Marshall, Thomas R. Jr. (petroleum geologist and former selection officer of Alaska Division of Lands, Dept. of Nat. Resources), 90
McCarthy Town, 33, 35, 41, 65, 90, 138
Medley, Edward F. (Seattle lawyer for Whitham), 68
Mentasta Mountains , 32
Mentasta Village, 20, 49
Merchant Exchange Quarters (Seattle), 64
Miles, Nelson A. General, (ordered Lt. Allen to conduct reconnaissance of Interior Alaska), 24, 27
Miles City, Montana, 24
Miles Glacier, 27
Mint, United States (all Nabesna gold sold to U.S. Mint), 96, 98-99
Moffit, Dr. Fred (early day USGS geologist who mapped Eastern Alaska Range), 34, 116, 138
Morgan Bank (helped finance Kennecott Mines), 65

Morrison-Knudsen Co. (also MK and built Reeve Field), 120
Mother Lode Copper Mine, 33, 37, 39, 65-67
Mother Lode Coalition Mines Co. (owned by Kennecott), 65, 67
Muller, John, 68
Mt. Sanford, 15-16, 26-28, 32

Nabesna Mine, 1, 10-11, 13, 15-16, 24, 26, 34-36, 46-47, 49-51, 67-68, 84, 90-91, 99, 101, 103-106, 115-124, 126, 129, 133, 152
    100 level also Bear Vein drift, 61-62, 63, 89, 114
    250 portal and level, 11, 78, 80, 84-86, 88, 109, 114
    350 level, 84
    450 level, 84
    550 level, 84
    650 portal and level, 12, 84-86, 88, 111, 115
    Nugget Tunnel, 85-86, 88-89, 143

Nabesna Mill, 14, 93, 100
    Aerial tramway, 7, 71-72
    Assay laboratory, 95, 97
    Jaw Crusher, 91
    Ball mill, 91, 93, 96, 141
    Concentrating Tables, 93-94, 96
    Flotation, 96, 98, 142
    Cyanidation, 97-98
    Power plant, 98, 100, 102
    Mill tailings, 93, 100, 121

Nabesna Town, 15
    Post office, 109-110
    Store, 101
    Medical dispensary, 101
    Water supply, 15
    Mess hall, 90, 101-102, 109, 113
    Bunkhouses, 101-102
    Do Drop Inn, 110-113

Nabesna Fraction (reported as CIC underground cell at Nabesna during Cold War), 157
Nabesna Glacier (reported as World's longest alpine glacier), 5, 8, 32, 105-106
Nabesna Gold Mine, 1, 5, 7-8, 25-26, 30, 51, 60-61, 71, 82, 100-101, 119
Nabesna Landing Field (built by Whitham in 1930 for Reeve and others to fly gold out), 1, 77, 79, 95, 103, 119, 120-122

Nabesna Mining Corporation (owned Nabesna Mine), 3, 21, 58-59, 61, 68, 70, 90, 116, 123, 131, 139
Nabesna River, 5, 8, 25-26, 31, 35-36, 39, 44-45, 58, 69, 77-78, 95, 119, 120, 122
Nabesna Road, 1-2, 5, 7, 14-15, 19-20, 24, 27, 29, 32, 36, 44-46, 50, 38, 60, 71, 79, 89, 95, 100, 102, 104-106, 111, 113, 115-118, 120, 133, 135, 138
Nabesna stock (granitic intrusion into limestone and source of metals at Nabesna Mine), 13-14, 124
New York Stock Exchange, 64, 67
New York Times, 45, 60, 66
Nome, 44, 49, 119
Northway airlift (reported as World's largest airlift to 1941), 121-122, 131
North Wrangell District (one name of Nabesna area before the 1930s), 44

Ogle, Jance Reeve (daughter of Bob Reeve), 122, 127, 139
Orange Hill (large pyrite-copper mineralized area 14 miles south of Nabesna where National Park Service built visitor cabin in 2001), 5-6, 8, 32-33, 50, 58, 121
Order L-208 (Order closing Nabesna and other gold mines during World War II), 123, 127
Oliver, Major (Engineering Officer for ARC), 44

Patterson, Dr John & wife EA (well known plastic surgeon in Fort Worth, Texas and Navy surgeon during World War II in the South Pacific), 106, 108, 118
Pilgrim, Earl R.(mining engineer and first Prof. of mining at Agricultural College & School of Mines now Univ. of Alaska Fairbanks), 80-81, 83, 89, 90, 139
Pollack Flying Service, 103
S.S. Portland, steamship, 29

Rearden, Jim (author of Alaska history books), 39, 121-122, 126
Reeve, Bob (famous Alaska pilot and flew gold from Nabesna), 78-79, 95, 103, 105, 114, 117-118, 120-122, 124, 131, 157
Reeve Aleutian Airways (formed by Reeve after World War II), 117, 122, 124, 127, 139
Reeve Field (originally called Nabesna Landing Field built by Whitham in 1930), 1, 78, 122-123
Renton, WA, 40, 43
Richardson, Wilds Captain then General (first president of ARC), 43-44, 49

Richardson Highway (First road between Valdez and Fairbanks and named for Gen. Richardson), 29, 39, 43-45, 49, 57, 60-61, 71-72, 79-80, 84, 89, 94, 100, 102
Roosevelt, President F.D., 49, 95, 104, 133
Royal Development Corp. (early mining co. at Nabesna), 31, 34, 38, 57
Royal Development Drift (old drift used by Nabesna fraction during Cold War?), 131
Russia and Russians, 1, 21, 23-24, 26-28, 34, 119, 122-123, 126, 129, 130-131

Sargent's camp, 32-33
Sargent, D.C (early mining man in Nabesna area), 32-33, 58, 110
Saturday Evening Post, 45
Seattle Chamber of Commerce, 45, 131
Seattle Stock Exchange , 63
Serbrinkoff, Lt. Ruffus (Russian explorer killed by Batzulneta Indians), 23, 25
Shushana (1912 gold rush town now called Chisana), 35-38, 66
Sheep Cave (at Nabesna), 21-22, 26, 51, 88
Sheppard, Ken (World War II colonel and well known Alaskan engineer and banker), 122-123, 127
Siberia, 19-20, 126
Slana River, 24, 43, 34-36, 39, 44, 50, 57, 104
Slana Roadhouse (early roadhouse built on Nabesna Road), 2, 45-47, 50, 61, 66, 72, 80, 89, 103-105, 115
Slana-Tok Cutoff (Glenn Highway), 2, 5, 10, 16, 29, 34, 39, 89
Slana, 5, 31, 40, 44-45, 50, 66, 89, 129
Skarland, Ivar (Prof. Univ. of Alaska), 113, 118
Skarn (geologic term for a type of mineralization at Nabesna), 13-14, 16-17, 22, 81, 83, 96, 142
Skagway, 29
Skookum Creek (crosses Nabesna Road near Devils Mt Lodge), 15, 111, 113
Skookum Creek caldera, 15-16
Skookum Creek Volcano. 11, 14-16
Smith, Merle "Mudhole", 124
Snow Gulch (placer claim at Shushana), 36
Soviet Union, 130
Sportsman Paradise Lodge (at Twin Lakes on Nabesna Road 15 miles from mine), 71
Stanford University, 104, 117
Steese, Jas G. Colonel (President of Alaska Road Commission 1924), 44, 54, 131

Standard, E.T., 44
Steward, Claude (Vice President of Nabesna Mining Co.), 123
St. Joseph Hospital, Fairbanks, 101, 115
St. Michael, 27
Stamp Mill Gulch (at Nabesna), 30-32, 88-89, 124
Stamp Mill Road (at Nabesna), 86, 88
Strandberg, Harold (President of Strandberg Mining Co. and once the 2nd largest gold placer mining company in Alaska), 127
Sutherland, Dan (Delegate to Congress and friend of Whitham), 43-44, 47, 49, 57, 60
Superior Court of State of Washington, 68
Supreme Court of New York, 68
Suslota Lake Pass, 26
Sullivan, George (knew Whitham and later became longest serving elected mayor of Anchorage, Alaska), 104, 117
Symington, Stuart, 130

Tacoma Smelter, 97-98, 107, 124-125
Tanada Creek (site of Batzulneta Village), 24
Tanana River , 20, 26-27, 34, 120
Taylor, Ike P. (Ass't Chief Engineer ARC 1927), 54
Taylor Highway, 29
Tebenkof, Capt. Michael (Governor of Russia-America 1845), 23, 27-28
Territorial Road Commission, 43-44, 49
Tetlin Lake, 120
Tetlin Village, 120
Thompson Pass, 84, 102, 117
Thornton, D.P., 37-38
Thornton, Marie (wife of Carl F. Whitham), 37-38
Tin house, 111, 113-114
Tok Junction (also Tok), 2, 89
Tok River, 29, 34
Tower, Elizabeth (author of many history books about the Kennecott era), 34, 44, 50, 69, 140
Twin Lakes (adjacent to Nabesna Road), 20, 49, 71, 91

University of Alaska Fairbanks, 62-63, 80, 106, 108, 118
University of Santa Clara, Calif. (Father Hubbard served as Prof. of geology), 106
University of Washington, Seattle (Holdsworth attended the Univ. while working at Nabesna), 28, 81-82, 86, 97-98
U.S. Army Signal Corps, 57, 103, 117
U.S. News and World Report, 130

Valdez, 27, 29-32, 34, 37, 39, 41, 43, 49, 57, 79, 84, 95, 98, 100, 102-10-107, 117-118, 120, 124
Valdez Chamber of Commerce, 106
Voss, Edna (Secretary of Presbyterian Chruch New York), 47

War Production Board (World War II agency), 123, 127
Wayland, R.G.(USGS geologist), 116, 118
Whitehorse, Yukon Territory, Canada, 119
White Mountain Cabin (first cabins at Nabesna 1898), 30-32, 39, 73, 88
White Pass, 29
Whitham, Carl F. (founder and president of Nabesna Mining Corp.), on page 1 and throughout text)
Whitham, Marie (wife of Carl F. Whitham), 39-40, 42-43, 45, 53, 57, 66, 71
Whittier, 119
Wickersham, James (U.S. Judge and elected delegate to Congress 1908), 28, 41, 60, 118, 137, 140
Windsor Hotel (Cordova), 39
Wollastonite, 13, 17, 22
World War I, 38, 49, 66, 90, 117
World War II, 1, 2, 17, 46, 50, 77-78, 89-90, 103-104, 105, 115, 117-119, 122-127, 129-131,
Wrangell, Admiral Baron F. von (Governor of Russia-America and namesake of Wrangell Mountains), 23, 28
Wrangell-St. Elias National Park and Preserve (Est. 1980), 2, 6, 17, 21, 34, 100, 138-139
Wrangellia (name for a tectonic plate), 9-12
Wood, John W. (Board of Missions Episcopal Church), 49

Yale University, 45, 127
Yukon River. 24, 27, 29, 31, 34, 50, 57, 103

Zenith (one of Gillam's airplanes), 105